This System is Killing Us

"Dunlap is one of the foremost researchers on the unfolding relationship between ecocide, colonialism, extractivism, and green capitalism. The reason he is able to unmask the realities of the supposed solutions to the ecological crisis—profitable platitudes like 'green energy'—unlike so many professional academics who continue to dilute their critiques or even promote the very activities that are pushing us over the brink, is that he has cast his lot with the communities and movements that are fighting for our collective survival. This book is an important new contribution to his work."

—Peter Gelderloos, author of *The Solutions are Already Here*

"Dunlap's work is vital for understanding the forces driving violence against land and water defenders around the world—and why a transition to 'renewable' energy will fail to stop it."

—Alleen Brown, investigative journalist

"This book does what many of us cannot—it communicates truths about our world that we have instinctively known since we were children, but never been able to articulate. Dunlap illuminates the state apparatus, its various forms of oppressive tactics against the life that it depends on and its criminalization of land defense in a way that is thought provoking and at the same time accessible and understandable. If, like me, you like arguing with your family around the dinner table, this book is going to be your greatest accomplice."

—Bojana Novakovi, actress, film maker, and organizer

"Weaves together interview testimonies, fieldwork observations, and groundbreaking investigative research to offer a richly evidenced and passionate account of the harms of green capitalism. An indispensable resource for scholars, activists, and policymakers looking to make real change toward a socially just and sustainable future."

—Anna Feigenbaum, author of *Tear Gas: From the Battlefields of WWI to the Streets of Today*

"This superb book is a breath of fresh air amid established modernist and leftist treatments of the polycrisis."

—Arturo Escobar, author of *Pluriversal Politics*

"One of the most compelling, demystifying, and provocative calls to action in the face of the violent collapse of modernity. A must-read for anyone who wants to carry out or support serious anti-colonial, anti-state, and anti-capitalist struggles."

—Dr. Carlos Tornel, author of *Gustavo Esteva: Life and Work of a Public Deprofessionalized Intellectual*

"Demolishes our complacent faith in renewable energy and its associated fantasies of 'green new deals' and 'sustainable development', confronting green capitalism with the resistances its violence inspires, and drawing compelling lessons for our collective survival."

—Dr. Japhy Wilson, Lecturer in Human-Environment Interactions, Bangor University, UK

"Traveling across sites of ecocide and resistance, from Mexico to Portugal, this book investigates socioecological catastrophe and social war. But the critique goes so much further and deeper—exploring the links between the fabrication of desires, the enchantment of modernity, and the ongoing infrastructural colonization through statism, 'green projects' and the deeper pathology of progress. A simultaneously depressing and empowering must-read."

—Dr. Andrea Brock, lecturer in International Relations, University of Sussex

"Our modern infrastructural systems are killing us, and the planet. In this crucial book, Xander Dunlap exposes the deadly connivence of states and capital, the dangers of well-intended critiques, and the challenging realities of vital socio-environmental struggles."

—Professor Philippe Le Billon, Department of Geography, University of British Columbia

"Dunlap brilliantly documents the catastrophic outcomes locally and globally as we entrust our planetary woes to the promises of dapper executives, left or right, who peddle technological transitions for green growth. Why must we instead encourage each other to shift our ways to address poverty without growth? Answers inside."

—Professor James Fairhead, Department of Anthropology, University of Sussex

This System is Killing Us

Land Grabbing, the Green Economy and Ecological Conflict

Xander Dunlap

First published 2024 by Pluto Press
New Wing, Somerset House, Strand, London WC2R 1LA
and Pluto Press, Inc.
1930 Village Center Circle, 3-834, Las Vegas, NV 89134

www.plutobooks.com

British Library Cataloguing in Publication Data
A catalogue record for this book is available from the British Library

ISBN 978 0 7453 4882 7 Paperback
ISBN 978 0 7453 4883 4 PDF
ISBN 978 0 7453 4884 1 EPUB

This book is printed on paper suitable for recycling and made from fully managed and sustained forest sources. Logging, pulping and manufacturing processes are expected to conform to the environmental standards of the country of origin.

Typeset by Stanford DTP Services, Northampton, England

Simultaneously printed in the United Kingdom and United States of America

This book is dedicated to all the people struggling to defend their habitats and lives, and to stop the spread of toxic materials and relationships—everywhere. This includes the imprisoned land defenders in Atlanta, and elsewhere, suffering imprisonment and (false) domestic terrorism charges for defending trees, animals, and rivers against a police urban warfare training facility.

What concentration camp manager, national executioner or torturer is not a descendant of oppressed people?

—Fredy Perlman

The road to hell is paved with good intentions.

—proverb

Contents

Figures

Prologue

I am still haunted by the interviews I conducted between January and March 2018 in southwest Peru. This, by no means, was the first time I experienced or confronted testimonies of political violence, as this book (and others) reveal.[1] After an invitation from my friend, Carlo, to examine the socioecological impacts of the contested—and still not operating—Tía Maria copper mine, which residents of the Valle de Tambo (Tambo Valley) have been opposing since 2009 (see Chapter 4). By the time I arrived in this region, there had already been a decade of contestation, which included two police and military invasions. The first in 2011 and the second in 2015, the latter becoming a State of Emergency that resulted in a sixty-day military occupation. The Peruvian state sought to put an end to the (defended) barricades and general strike, which was preventing Southern Copper Peru from beginning its mining operations.

At this point, Carlo had arranged interviews for me but I was also going door-to-door and interviewing people on the street. The testimonies, mostly from women, were saturated with tears as they remembered police house raids in the early hours of the morning. Police breaking into houses in tactical gear, dragging people out of their beds, with naked children crying in the street, and abducting people suspected of making barricades, slowing or thwarting police invasion (Espartambos).[2] The pain and emotions I stirred, conjuring in people the feelings of violence, helplessness, and anger they felt, by trying to understand the impact of a copper mine was contagious and cause for critical self-reflection.[3] These conversations are branded in my head, creating a foundation for when I met "Sally."

Waiting 20 minutes on their front steps, in a small hamlet on the side of a hill, I watch Sally approach. The walk of caution, the look of suspicion and the intensity in the eye, indicating trauma, was memorable. Despite this, Sally was kind. Inviting me inside their home, bringing me water and graciously sharing with me their experiences with the struggle against the copper mine. While going over the standard script: "How long have you lived in the valley … What do you think about the Tía Maria mine? … How did you experience the State of Emergency?" Sally told, Carlo and myself, how this mine has affected her life. She talked about how she left her father in Callahuanca:

> Sally: Because my husband has to work and I have to come back down into the [Tambo] Valley with him. So my father told me, "I am going to stay here, but tomorrow I go down." So the next day he called me and he said, "Daughter I am coming down to you." And I said, "Yes, daddy, come down and I will have breakfast ready …" So the next day he makes me cook the food that he loves and I said, "Dad, why are you asking for so much meat?" [He said:] "Daughter, cook it for me, cook it for breakfast and prepare it as you know how to, barbecue it." "Yes dad," I said, and it made me think.
>
> I woke up very early the next day at four or five in the morning, I was cooking very early and I prepared what he asked for breakfast, and he ate everything. Then I asked my father: "Daddy are you going to the strike today? Because today the strike is going to be hard." Because that day I remember that Arequipa [the regional city] is going to stand up and join the strike. Arequipa is going to support us. My dad told me, "Arequipa is going to support us, so today, yes, we are going to cross [the bridge]." What the Tambo Valley wants is to cross the [Pampa Blanca] bridge in Chucarapi. The police were on both sides and they would not allow people to cross, so my dad told me: "So yes, today because Arequipa is going to stand with us, we are going to cross to arrive in El Fiscal to protest." "But daddy, it is going to be heavy," I told him, "but you do not need to go. It's better to go see the plants." And he told me, "Maybe I will go to our plot of land." And he was like this, he was unsure whether to go to the strike or the field.
>
> So the thing is I went to La Curva to help weigh peppers with my father-in-law and I came back here at 1 pm or 1:30 pm and

I did not know where my father was, he was not here. I thought to myself: "Maybe he went to the strike? Where is he?" So I also started going to the strike at 1 pm, but my son said, "Where are you going to go mother?" I said, "I am going to the strike for a while." And he told me, "Don't go." So I stayed in the house for a while, washing and doing stuff until people called me. "Sally!" "What happened?" I said. "Your father is bad," they told me. I said, "Why is he bad? Where is he?" [The person on the phone explains:] "He is here in the strike and he got shot." At that moment I thought that it was birdshot. I did not think it was a real bullet. "It's a birdshot bullet." [Crying]. "It's birdshot," I said, "he is going to be okay?"

For me, my dad was untouchable because he was both our father and mother for the family because we were orphans since we were little. "I do not think he will be okay," I said. The other person said, "Yes in the foot. It must be birdshot; I do not think it's a real bullet." And then, more people called me and told me, "Sally also your brother." And I said: "What?! My brother and my father!?" At that moment I got crazy on the highway. These people make mistakes, it was only my father—my dad! So I ran to Cocachacra, I went crazy because there were no cars on the highway [because of the general strike]. I had no idea how to get there, but I had to get there. When I arrived at Cocachacra, they were no longer there. When I was going down the road, I crossed an ambulance. So my father was leaving in that ambulance and when I went back [the other direction] and I arrived in Mollendo the nurses tried to calm me down: "He is going to be okay, in a moment he will come out [of the room]." At that moment, I found out it was not birdshot. He was shot with a real bullet; they had shot him in the butt from behind and it went out the front. I thought it was birdshot, but then the doctors came out and told me, "He's gone." [crying]. He's not here, He's not here, he is not [crying, crying]!

[Silence]

AD: I am sorry, really. The state is rotten.

[Time passes]

AD: Was there ever a judicial process about this? Some process?

Sally: Yes, my brother filed a complaint with Dr. Hugo Herrera and I did it with the [Arequipa] School of Lawyers … Sometimes I search for Dr. Hugo Herrera and I ask, "What about the case of my father?" He says, "We are investigating," just that—investigation. And Dr. Vildoso from the School of Arequipa who was mainly with me, he told me: "No, the case is closed for a lack of evidence and I do not know." "What evidence?" I said, "There are a lot of videos, how many videos do they need? There are a lot of recordings!" I said, "How could you tell me there is a lack of evidence?" I think this is because of a lack of money, not of a lack of evidence. Regrettably, I am poor and I cannot contribute with one Sole, so I have nothing, what can I do? For a dog there is justice, when there is a video that he was killed, there is justice. So my father, what is he? Not even a dog? So, what can I do? My husband also told me to "let it go, it's making you sick. It's making you feel bad."

So there are a lot of things that happened to me. My family was almost at the point of destruction (and my brothers do not know); I went mad searching for justice and guilt to the point I became violent against my husband, even one time I took the knife wanting to hurt my husband, because my husband would say to me: "What are you going to do? The Company is like this." Then my mind only wanted to hear things that agreed with it, for the pain maybe. And like this, what can I do? There is no justice for my father.

Once we turned off the voice recorder, she began to explain that when her father died, life turned into a "living hell." Sally, her husband, and father took out a loan of 15,000 soles (approx. 4,617 USD) from the bank to share a plot of land. Both Sally and her husband had jobs so they paid her father to work, but then he was killed. This led to severe depression for Sally. Then the State of Emergency was declared and she had to take care of two children while her husband had to travel to Cusco for work. Months passed, nobody could work the farm, and the bank debt grew to 60,000 soles (approx. 18,468 USD). She was all alone, depressed and started to drink and her children did not want to be around her because she would cry all day and, in moments, wanted to kill herself. She would spend all day at the cemetery with her father's grave, even sleeping there sometimes. She began attending therapy, talking and working in the field, which earned her debt relief. When we talked to Sally, her life was slowly improving, step by step, and aided by working the fields that were threatened by the mine that would compete with agriculture for water sources.

Sally's story and others' in the Tambo Valley and elsewhere discussed in this book still haunt me. This story, the pain communicated through apprehensive body language, the intensity and pain pouring from her eyes still reverberates through my body. These memories, or "psycho-social ghosts," are antagonized frequently and are conjured by corporate environments. From universities to hipster cafes, I am repeatedly in situations where well-dressed people, reeking of confidence, are expounding with an ambiguous enthusiasm (and implicit hope) about how "renewable energy," "energy transition," and XXXX new technology will prevent "climate change," as if it is not already happening. The word "entrepreneurship" is not far behind, meanwhile covert racism and "ontological-epistemic" discrimination[4] begins to show its face in these conversations, or lurks ready to leap out to discredit the experiences challenging Western materialism and market democracies. The uncritical embrace of capitalism, industrial technologies, and governance reminds me of the (relatively) free-range cows trotting around carelessly stuffing their face full of dry grass and brambles on the mountain behind Oaxaca City—pure of heart, careless, and secure in their place. To be sure, a carelessness naiveté to envy but nonetheless disrespectful as we sit playing "devil's advocate," mediated by a table, reliant on central heating and postulating about how to "reduce carbon dioxide" to "save the planet," as if we are an elite policy maker or Leviathan incarnate. The violence of these civilized or university environments conjures Sally's story, the tears of friends and strangers along with the violence of street fighting in my head as the living and dead ghosts of modernity weigh heavy. This pain sits with me, it remains an underlying motivation for writing this book and, "yes," I want you to meet these people and ghosts. I think it is necessary if we are going to talk politics, or even imagine political action.

What comes in the chapters that follow is a testament to the incompetence of governments to address socioecological and, consequently, climate catastrophe. This might be obvious for many[5] given the pronouncements of climate marches and activists setting themselves on fire.[6] There is a lot taken for granted knowledge and actions in terms of how socioecological and climate change *will actually be slowed, mitigated, and stabilized.* More common right now are people "demanding" that governments fix climate change, uncritically leaping into hopeless green capitalist schemes, or imagine seizing the state to institute an authoritarian "socialist" or ecological Leninist regime, as if people have not been trying to "change the system from within" the entire time. This includes the necessary, if late, rise of degrowth as a popular topic within universities as it slowly makes its way into policy circles. This book contributes to this conversation. Assuming the goal is to stop socioecological harm and regenerate relationships with our habitats, the following main chapters document the socioecological impacts—if not horrors—generated by technological development and its green capitalist solutions to "mitigate" climate change. Between green capitalism(s), reductive science—or arithmetic elevated to the status of science—and authoritarian desires, this book seeks to ground these conversations into the realities that beset us. This means, thinking of Sally and others in the book, examining autonomous struggles and their fight against the encroachment of "modernity," "technological innovation," and the destruction of their ecosystems. With all kinds of actors preying on the anxieties of socioeconomic hardship and ecological collapse, this book seeks to refuse the priests, politicians and the used car salesmen of the day—as charming and professional as they are. What follows is a reminder. Sally's experience, for example, is not what comes to mind when we think about copper wiring, wind turbine generators and the means to transport electricity (e.g. high-tension power lines). It is time to include these realities and understand why people are defending their territories. Undoubtedly, confronting the state and capital creates an enormous challenge, but one we must face honestly with our communities and ourselves if there will be any socioecological transformation toward regeneration and real socioecological reciprocity. Understanding the political depths, complications or long-term nature of socioecological struggles, the book shows that only "we," "us" and "everyone" will be the ones who will have to change, fight and struggle. Supporting each other from wherever we are, according to our skills and abilities to subvert the socio-ecological catastrophe that is well in process, should remain the highest priority—and really concerns everyone.

Introduction

All political conversations are rooted in what we need, think we need, or want. If our priority is to have information technologies—smartphones, video games, and entertainment industries to name a few—then this will condition our desires, political imagination, and real or imagined developmental trajectories. While I might be addicted to information technologies—checking my emails, Twitter, or cartoons (e.g. *Rick and Morty*, *South Park*, or *One Piece*), even if the former is a habit acquired through work and academic industry[1]—I am consciously unattached to this lifestyle or political trajectory. In fact, I work to break with them. This lifestyle, and its corresponding infrastructures (e.g. paved roads, mobility regimes, networked water, and electrical systems), demand an enormous number of raw materials and energy from the earth and its inhabitants. Despite all the marketing and scientific gymnastics employed to suggest otherwise, this is why industrialized capitalist systems are completely ecologically unsustainable. Equally important, I realize my favorite shows or computational pastimes are just a "fix" or substitutes for other meaningful activities I could have in my life.[2] That is to say, I enjoy the fruits of modernity while they are here, but I am not attached to them, fighting for them, and at the least I am striving to find a healthy balance. I recognize that this lifestyle is ultimately slowly destroying the earth, and that my addictive romance with technology comes at a cost to other relationships and pastimes and, like the earth, is degrading my life. The blue light from computers and screens, Rupa Marya and Raj Patel remind us, causes disruptions in circadian sleep rhythms that "lead to an increase in inflammation-driven diseases such as diabetes, depression, high blood pressure, and obesity."[3] The profound, rippling, and accumulative impacts of modern technologies should not be underestimated, nor ignored.

Even if I could obtain a balance between work and/or technological pastimes, it would not change the uncomfortable fact—which governments, companies, and people have been avoiding at all costs—that modernist, consumerist, or techno-capitalist lifestyles are a process of spreading industrialized (e.g. highly processed) material and toxic waste across lands, forests, rivers, marshes, and, ultimately, the planet. Most of the industrial materials surrounding us and that we inhabit—steel reinforced concrete, drywall, particleboard, and even still asbestos—are either forms of low-level or high-level toxic materials. Municipal water, if not also having chlorine in it, actively adds fluoride—a fertilizer industry's chemical by-product—that recent studies show actually causes "neurological changes in children exposed to fluoridated water."[4] The "average city dweller," moreover, "breathes in 10,000 liters (2,600 gallons) of air a day, air tainted with tobacco smoke, automobile exhaust, diesel soot, ozone, sulfur dioxide, nitrogen dioxide, and a smattering of other pollutants."[5] Tailpipe exhaust fumes, Marya and Patel show, "have been linked to a higher prevalence of brain cancer, so much so that the elevated exposure that would occur if you moved from a quiet street to a busy street, and stayed there for a year breathing the air, would increase your risk of developing the disease by 10 percent."[6] There are many ways to inhabit and build within our ecosystems, but the industrial human—in all of its great technological feats—spreads toxins and poisons in their environment without hesitation, with little consideration and, even while knowing its consequences, continually fails to stop these patterns. For some time now, people have been violating the number one rule of habitation: *Do not shit where you eat (and sleep).*

This book is about the reality of our modern infrastructural systems, be they conventional mining operations or so-called renewable energy. As I have shown elsewhere[7] and will demonstrate in the following chapters, I say "so-called renewable energy" because wind turbines, solar panels, and dams are *not renewable* as we currently know them. Specifically, how they are manufactured, designed, and operated within capitalist systems seeking to accumulate more energy, profit, and power. While this will be a recurring thread throughout the book, said quickly, what governments and companies market as renewable energy is in fact an entire system of large-scale mining, mineral processing, manufacturing, digital applications, transportation, and (often failed) decommissioning processes, which are all powered by fossil fuels, natural gas, and continues large-scale ecological degradation. The issue of accounting, which justifies robust claims regarding renewables, will be the focus of the next chapter. This criticism should not be confused with not desiring or supporting the creation of genuine renewable energy, because modernized humanity *needs to learn how to create renewable energy systems that organize ecological regeneration and socioecological infrastructural sustainability*. Because, as many readers should already know, the present socioecological situation is critical: 40% of the planet's soils are seriously degraded; earthworm biomass has declined by 83%; global fish stocks facing 85% depletion; mammal populations have dropped by half; and dead zones from chemical run-off, nitrogen, and phosphorous spread along the coastlines of industrialized regions across the world.[8] Deforestation, furthermore, is at an all-time high, as forest fires, record heatwaves (resulting in human and nonhuman death), erratic weather and floods spread across the world with rising sea levels.[9] Meanwhile, "[c]ancer rates are 2.5 times higher in what we call 'developed' nations than in poorer countries," and while "poor countries push for so-called development, their cancer rates are rising too, even when controlling for things like improved detection and population aging."[10] It is time for people to get serious about ecological problems, which are social problems in terms of governance, social organization and relationships that will produce and maintain the present statist and capitalist catastrophe. Hence, the reoccurring use of the term "socioecological," because the two are inseparable. This means engaging in *permanent ecological conflict* until ecosystems are re-inhabited, and protected and socioecological crises reconciled to begin a process of environmental resurgence. This book examines numerous environmental conflicts, peering into different groups and people who are committed to engaging in permanent ecological conflict. Capitalist political economy must be transmuted and reconfigured as past cultures and ecosystems once were to serve state formation, empire, and Industrialization.

Ecological and climate catastrophe, discrimination against different ethnicities, peoples, and sexualities as well as endemic police violence and war remain clear indicators of socioecological instability and crisis, even if capitalist economies tend to grow and thrive under these conditions. This book seeks to put this reality into perspective, decentering the existing capitalist delusions related to geoengineering and various green authoritarian propositions from the right and the left that suggest a stronger state can regulate and resolve socioecological catastrophe—a theme revisited briefly in the next sections. This means challenging easy or familiar statist, capitalist, and modernist solutions by relying on *ourselves and each other*—wherever we are—to appropriate and transform existing technologies and bureaucracies to create genuinely renewable socio-infrastructural systems and habitats. While this is a daunting task, if we do not accomplish this then the present trajectory of socioecological catastrophe will continue and intensify. This means, thinking of Alfredo M. Bonanno,[11] entering a process of *permanent ecological conflict* that—from each according to their skills, desires, and capabilities—begins a process to stop the forces of socioecological domination and degradation, while organizing integral habitat regeneration and renewability in the widest sense.

The first chapter discusses knowledge production and environmental metrics to frame serious cultural-scientific problems related to the perpetuation of socioecological catastrophe. The remainder of the book will explore different methods of ecosystem extraction. This includes wind energy development in Oaxaca, Mexico (Chapter 2), coal mining in central-west Germany (Chapter 3), copper mining in southwest Peru (Chapter 4), low-carbon energy infrastructure development in southwestern France and southern Iberia (Chapter 5), and, finally, lithium mining in northcentral Portugal (Chapter 6). Each chapter will delve into the realities and complications of these ecologically extractive and infrastructural activities, meanwhile discussing the resistance struggles that accompanied all of them. Besides conducting various durations of fieldwork in these sites with friends, what ties all these environmental conflicts together is peoples' willingness to fight—or begin learning how to fight—and defend their habitats from extractive encroachment by statist and capitalist forces. The conclusion will reflect on these efforts and conceptualize ways to proceed based on the experiences that arose from the research in these chapters.

The desire to resist this catastrophe, we must recognize, is strong. There is popular acknowledgment of socioecological catastrophe and climate change with glaciers melting, and renewed interest in resistance from a new generation of people symbolized by Greta Thunberg's entrance into the media and conference spotlight. This manifests with Fridays for Future protests, Extinction Rebellion's (re)popularizing of civil disobedience actions and the overall mainstreaming of the climate justice movement. Meanwhile, academics have jumped onto this bandwagon promoting different strategies of socioecological transformation, from various articulations of the social democratic Green New Deal, "green" state dictatorships and degrowth. While these proposals receive mainstream attention, autonomous, anarchist and Indigenous struggles—as one or separate—continue struggling against gas pipelines in Canada,[12] anti-police uprisings across the Americas, and an insurrection in Chile triggered by the increase in transportation prices, which brought more issues to the surface and almost toppled the regime. Meanwhile, the anti-colonial Palestinian struggle continues beneath tanks and drone strikes, Yemenis bombardments by Saudi Arabia and the Russian invasion of Ukraine also combine among countless other land defense struggles across Latin America, Africa, and Asia. Political turbulence, struggle and war rage on while the ecological conditions foreseen since the 1950s are worsening[13]—and no thanks to imperial powers prolonging their ecocidal war efforts.[14] This book retains a narrow focus on land struggles, mostly focused at the regional level within countries to reveal the techniques of land control and (internal) colonization that are frequently taken for granted.[15] This book delves into how governments and companies—as they are often colluding—divide, conquer, and control humans, but also the existence of rivers, trees, fields, and animals. This book is an exploration into the mechanics of territorial control and (re)colonization of regions by governments, companies, and their infrastructures. The remaining sections in this chapter frame the recent popular academic debates this book contributes, the theory guiding this work and the methodologies behind the research in the following chapters. Readers, uninterested in popular academic debates, theory, and methodology, are encouraged to turn now to Chapter 1. The next section, however, peers into the troublesome popular academic debates and the discontent they have sowed.

SHOCKED BY THE ANTHROPOCENE: DEGROWTH, THE GREEN NEW DEAL, AND MEGALOMANIA

The way some researchers confront socioecological catastrophe is to label these times the Anthropocene. Coming from ancient Greek, *Anthropo* means "human" and -cene from *kainos* means "new" or "recent," the Anthropocene comes to mean the New Human Epoch or the Age of Man.[16] This highly debated concept indicates that humans are the greatest influence on the planet because they are radically changing the ecology and topology of the earth. While *modernist humans* and the compliance of people with particular governance, infrastructural, and consumption regimes are having an undeniable influence on the planet, the traditional framing of the Anthropocene tends to homogenize all of humanity as responsible for the unfolding socioecological and climate catastrophe today. This framing forgets colonialism, slavery, capitalism, the state systems, and the mode of production and specific actors and technologies that created the present state of the world. The Anthropocene, likewise, has been interpreted as a narcissistic framing of Man "even proclaimed himself a 'geological force'" and the center of the universe, confessing "his guilt for the unprecedented extinction of plant and animal species."[17] The Anthropocene is the most flagrant expression of Western humanism, which remains a propulsive emblem of the present catastrophic situation.

This has led to an overwhelming proliferation of terms to counter this proposed renaming of the geological epic. To list only a few, Jason Moore famously responded with the Capitalocene, nodding to Karl Marx's analysis of "capital" and political economy. The Capitalocene pushes for accuracy and locating culpability for ecological destruction by pointing to capitalists and capitalism.[18] Equally powerful has been the Plantationocene, which speaks to colonial ventures and the system of slavery (Negrocene) designed around the plantation economy. The plantation economy, thus, broke matrical bonds between numerous peoples, mostly Black and Brown, and the earth, wrecking socioecological relationships of care and reciprocity in the service of colonial (and statist) ecology. The fruits of which bloom with the (fascistic) racism, xenophobia, patriarchy, and ecosystemic collapse small-and-large.[19] The Necrocene (and/or Thanatocene) or "New Death", moreover, "reframes the history of capitalism's expansion through the process of becoming extinction," which emphasizes the death created by colonialism, capitalism, and present technological and economic developmental trajectory that has planetary repercussions.[20] While the Anthropocene is a technocratic tool—that seeks to make it seem technical, neutral, and legitimate—to make palatable the destructive impact of extractive development, it also serves as a governance tool and a psycho-social manipulation—for how we conceptualize the planet and our activities within it. "Saying humans are responsible for ecological devastation is a continuation of colonial racism, and it is an insult to the peoples who have fought against obliteration to preserve their way of life and their relationships with their territory," explains Peter Gelderloos. "It is also an insult to the many people who, despite growing up in a culture totally infused with the values of capitalism, have risked their lives and freedom to defend the land and halt destructive development projects."[21] It is from this perspective, I would suggest, that everyone should discontinue using the term Anthropocene, regardless of peoples' investment or attachment to the term. This term seeks to describe, homogenize, and essentially attempt to discursively manage the world in one term, while promoting the present trajectory of ecological modernism and hyperactive green capitalist development.

This inclination toward control, rooted deep in Western and civilized cultures,[22] inserts itself into every conversation when people are confronted with uncomfortable political realities (which people are rather happy to ignore in daily life). The question, often with righteous indignation, presents itself: "Well … What is the solution? You always talk about what is 'wrong' with everything, but what is the solution?" Solutions, we must admit, are about control. How to control the situation, make the problem and negative emotions go away so people can implicitly go on about their lives: working, getting intoxicated, playing with social media, and what have you. This "going away," or reactive revulsion, is the issue. People want others to deal with their immediate or distant problems for them. This relation conjures expressions of patriarchy with which we are all too familiar, where the male head of the household is responsible for making decisions and fixing the practical—but of course political—problems. There are many different types of patriarchy,[23] but gendered hierarchy and divisions of labor remain fundamental features. This manifests in the generalized expressions of the never-ending marketing of the used car salesman, and its political articulation as the politician who will tell you what you want to hear to gather a flock of followers to build a political constituency and, consequently, gain a managerial position within the state apparatus. From the household to politics, people want to believe in others doing the right thing, having the right science and moral fortitude to delegate fixing social, ecological and climatic problems—so they do not have to and can "live their lives." *If* people want to repair their social and ecological relationships, helping to create a path toward real renewability and beginning a genuine "climate change mitigation" process, then I am not sure if this delegation of labor and people relinquishing their power to authorities or companies can continue. This should alert readers to the issue of divisions of labor—instrumental to bureaucracy and complex mechanical organization—that is the intimate fabric of colonialism and the technics of the Necrocene. Remembering Ivan Illich,[24] people cannot keep putting their power uncritically into the hands of politicians, experts, and technological interfaces (e.g. internet, apps, social media), expecting them to fix things these very processes are enabling in one way or another. Instead, we have to learn to make our own decisions, organize consequences for life-destroyers, and take our own actions, individually but ideally as communities, neighborhoods, and bioregions.

When it comes to conversations on socioecological solutions, the Green New Deal, degrowth, and authoritarian leftism are some of the alternative solutions debated right now. The degrowth school, while containing multiple and differing voices,[25] can all agree that in order to avert socioecological catastrophe, a planned reduction of energy and resource throughput must be organized until the economy is back in "balance with the living world in a way that reduces inequality and improves human well-being."[26] The expansive tendencies of capitalism—transforming the planet into urbanized environments that produce toxic and nuclear wastes—consume labor, hydrocarbon, mineral, timber, and kinetic energy resources is placed front and center in the degrowth analysis. A key strength of degrowth is its focus on reducing material throughput—the "taking" and "grabbing"—which positions it, in the words of Corinna Burkhart and colleagues, as "the most radical rejection of the eco-modernist mainstream of growth-centredness, extractivism and industrialism."[27] Degrowth confronts the dominant myths of ecological modernism and "green growth," which believe that technological solutions (e.g. low-carbon infrastructures, carbon capture storage, nuclear power, and geoengineering) can remediate climate change and socioecological degradation while maintaining economic growth as we know it.[28] Degrowth, said differently, responds to the eco-modernist world that industrial humanity unevenly inhabits and is continuing to spread under various forms of statism and capitalism.

There are, however, various eco-modernist positions, which believe in command-and-control state administration of large-scale technological projects and economy, while others believe that capitalism and market mechanisms can correct ecological degradation by *decoupling* economic growth, and technological development from ecological degradation or, really, ecocide. Decoupling, contends, that the economy can grow, while ecological degradation can decrease. This, of course, will be accomplished by improving the efficiency of household and industry energy use; increasing wind, solar, tidal wave, hydrological, and biomass power generation (and high-voltage powerlines); electrifying transportation with lithium-ion batteries; decarbonizing mining operations with lower-carbon power generation, digitalizing mine controls, automating machinery, and integrating electric vehicles; retrofitting buildings and employing smart censor to monitor (and price consumption) in cities; intensifying monocrop or "climate smart-agriculture," and so on.[29] These "decarbonization" measures, moreover, would be integrated into the military and police as well.[30] The Green New Deal (GND), or European Green Deal (EGD) remains an expression of eco-modernism,[31] continuing the existing modernist, capitalist, or state-capitalist trajectories. It should be acknowledged that many eco-modernists might argue the state is not doing enough with geoengineering, nuclear development, increasing urban densities (while creating areas without humans in nature), and investing in technological innovation.[32]

Decoupling, and the "green growth" position, however, has been thoroughly discredited at length by ecological economists and degrowthers.[33] After conducting an extensive review of the green growth research, Jason Hickel and Giorgos Kallis, for example, conclude that the "empirical evidence does not support the theory of green growth." This is because, first, "[g]reen growth requires that we achieve permanent absolute decoupling of resource use from GDP [Gross Domestic Product]. Empirical projections show no absolute decoupling at a global scale, even under highly optimistic conditions." Attempting to be supportive, Hickel and Kallis do note that "some models show that absolute decoupling may be achieved in high-income nations under highly optimistic conditions," but, in the end, find "that it is not possible to sustain this trajectory in the long term." Second, green growth demands that industrial humanity achieves "permanent absolute decoupling of carbon emissions from GDP, and at a rate rapid enough to prevent us from exceeding the carbon budget for 1.5°C or 2°C." And, so far, they note, "while it is technically possible to decouple in line with the carbon budget for 1.5°C or 2°C, empirical projections show that *this is unlikely to be achieved, even under highly optimistic conditions*" (emphasis added). Highlighting the non-existent empirical evidence to support green growth, or eco-modernist, policy, Hickel and Kallis conclude, that because green growth remains a theoretical possibility is no reason to design policy around it when the facts are pointing in the opposite direction."[34] Other studies are finding similar results. Reviewing 179 articles that contain evidence of decoupling, Vadén and colleagues conclude that "the empirical evidence on decoupling is thin" and "the evidence does not suggest that decoupling toward ecological sustainability is happening at a global (or even regional) scale." Vadén and colleagues continue that analysis of decoupling "needs to be supported by detailed and concrete plans of structural change that delineate how the future will be different from the past."[35] These findings, indeed, raise serious questions of legitimacy concerning international financial, governance and higher-learning institutions that ignore the reality of ecological modernism and the necessity of degrowing material and energy production/consumption. Said simply, the guiding ideologies driving industrial and environmental policy have, and continue, to fail the health of the planet.

Degrowth, as opposed to capitalist liberalism and eco-modernism, gets to the root of human exploitation and nonhuman extraction, questioning developmental modes requiring enormous amounts of raw materials and energy. This also includes critically reflecting on the productivist work regimes organized, whether liberal capitalist, state capitalist or otherwise. Degrowth, while retaining differing tendencies within it, seeks to create a public space for socioecological remediation and promotes a largely anti-authoritarian developmental pathway by advocating "degrowth values," such as autonomy, care, conviviality, equity and direct democracy.[36] Degrowth is the organized and planned reduction of energy and material consumption with the intention of improving the quality of people's lives by moving toward more convivial—grounded and socially developed—technologies and fulfilling lifeways rooted in community-supported agriculture, commoning land, cooperative economies, switching to localized renewable energy production, and political systems built around direct democracy and more.[37] While it has been acknowledged "that three-quarters of degrowth policy proposals were top-down with a national focus,"[38] it still represents, even if in need of greater elaboration, a popular pathway toward autonomous, feminist, democratic, and anarchistic approach to social development. This, of course, assumes that "top-down with a national focus" degrowthers, integrated into state and corporate institutions, will actively support and work toward creating unmediated political (e.g. liberated) spaces of experimentation, collective resurgence, and anarchistic renewal.[39] Degrowth, as you can imagine, is not without its critics, from ecological modernists to the authoritarian leftists chastising their failure to have a planned program or pronounced focus on the working class.[40] Likewise, there are sympathetic critiques from feminists and anarchists for the Global

North and South, pointing out the relevance of degrowth ideas and policies but also how degrowthers appear detached from political struggles (with middle-class positionalities) and failure to be clear about political strategy and action.[41] Degrowthers, however, are working through these criticisms,[42] which are compounded by the conflictive reality of capitalism and the state. This means charting a viable path toward social transformation and that degrowthers' "strategic orientation thus needs a strategy for [how to engage] the state."[43]

The issue of the state quickly leads to the hopes surrounding the Green New Deal in all of its variants. While readers might be more familiar with the Green New Deal as it spread across headlines in 2019, it was initially a term proposed by the infamous conservative economist and *New York Times* columnist Thomas Friedman in 2007. The GND refers to President Franklin D. Roosevelt's New Deal that responded to the Great Depression with social and economic reforms. In January 2019, congressional representative Alexandria Ocasio-Cortez and Senator Ed Markey proposed the GND in the United States Congress. While it failed to pass in the Senate, it created enormous enthusiasm for renewing public policy with a variety of energy, housing, agricultural, and industrial reforms. Numerous authors advocated the Green New Deal,[44] among them Noam Chomsky, and the program was further elaborated on by economist Robert Pollin.[45] Later, the GND was further developed under Bernie Sanders' presidential campaign,[46] while the European Commission began enacting the European Green Deal. Trade unions and non-governmental organizations also began articulating their proposals, only slightly departing from the original US proposal.

The GND and EGD remained "green growth" strategies that claimed to organize a (socio-technical) energy transition from fossil fuels to so-called renewable energy, all the while ignoring the amount of minerals, hydrocarbon resources, manufacturing, and transportation supply chains necessary to roll out low-carbon infrastructures. Bernie Sanders' fiery rhetoric against hydrocarbon industries did not account for this material reality of "achieving 100% renewable energy" in the United States[47]—or similar claims within Europe.[48] The "clean energy shift—wind, solar, hydrogen, and electricity systems," The World Bank report admits,[49] "are in fact significantly more material intensive in their composition than current traditional fossil-fuel-based energy supply systems." This issue will be discussed in the next chapter, but, in matters of political debate, people cannot forget that so-called renewables need, and are essentially, fossil fuels in terms of how raw materials are acquired, the chemicals used to extract ores and to operate manufacturing facilities.[50] No matter, the GND offered a valuable proposal to create "green jobs," agricultural reform, recognizing Indigenous rights, housing reform and promoting "just transitions" among others, which could have made incrementally positive social changes domestically and potentially by redirecting and restricting the use of hydrocarbons. Then again, unless the economic, energy and material growth imperatives of capitalism, and corresponding low-carbon infrastructures and electrification schemes, are capped—or have a limit—nothing structurally changes within this socio-technical shift that continues private or state capital accumulation. In the end—as usual—ecologies and habitats would be overlooked and sacrificed in the name of low-carbon infrastructures, green growth (or even a perverted version of degrowth), meanwhile (neo)colonial global supply chains predicated on unequal exchange, violence and racist discrimination would remain relatively untouched.[51] Not to forget, nobody really knows the quantity of fossil fuels actually used to produce a wind turbine, solar panel or dam. These issues are discussed further in the next chapter, with an emphasis on the arithmetic, models and science propelling these aspirations.

While degrowth advocates initially advocated the GND, seemingly uncritically—overlooking the realities discussed in this book[52]—it still led to heated and antagonistic debates with environmental economists and modernist socialists.[53] Despite the needed social reforms of the GND, the mainstream versions still never questioned economic growth, energy markets, and the expansive reality of capital accumulation responsible for socio-ecological catastrophe. Because, as James O'Connor reminds us, "over time, capital seeks to capitalize everything and everybody."[54] Said differently, capitalist systems seek to categorize, control and consume all life forms into some type of product to be bought, sold and traded. If the GND is anything like Roosevelt's New Deal, Gelderloos reminds us, then it is designed to prevent "a real solution" and "to save capitalism," placing "the brunt of this new industrial onslaught" onto the laps of the marginalized and poor of this world lower on the capitalist pyramid scheme.[55] From this perspective, the GND proposals sought to blunt revolutionary demands for socioecological transition, meanwhile developing and expanding green capitalism.

Implicit, and most appealing, about the GND is the state as an agent of administering social change. Experts, however, agree governments across the world, especially from Euro-America influencing international policy, have resolutely failed for 30—if not 40—years to develop adequate environmental policies that produce results.[56] Some people blame this on hydrocarbon companies lobbying politicians, hiding and falsifying science,[57] but this accepts the other deleterious socioecologically destructive impacts of urbanization, the proliferation of plastics, chemically intensive industrial production, and low-carbon infrastructures (dependent on fossil fuels) that are normalized by capitalist states in their quest for territorial control and technological supremacy. This fossil fuel versus renewable energy dichotomy emblematic of the GNDs, and inundating corporate propaganda, remains central to the "socialist modernist" position as well.

Similar to eco-modernism, the socialist modernist position takes on various intensities, yet has a core set of beliefs. "Softer" socialist modernist positions join the GND bandwagon, which celebrates centralized planning and technological innovation. "Solving climate change undoubtedly requires massive new industrial infrastructure in energy, public transit and housing," explains Huber.[58] This perspective, however, challenges capitalism based on a presumed ethic of egalitarianism and a pronounced concern with the "working class" and "global proletariat." This socialist modernist position celebrates and encourages "techno-fixes" such as carbon capture storage (CCS), nuclear power, and the state as administrator.[59] "Clearly, the productive forces must develop beyond their historically entrenched reliance upon fossil fuels," explains Huber.[60] This somehow implies, suggesting a reference to Marxian stage theory, that low-carbon infrastructures and electrification are separate from hydrocarbons and will usher in a new stage of decarbonized and renewable (socialist) industry. This tendency, moreover, tends to operate in the abstract with repeated references to Marxian theorists, for example, criticizing degrowthers for missing "the concrete class relationships that both inhabit such transformations or might bring them about." While Huber has been rebuked by other Marxian scholars,[61] it is strange how he failed to engage with Joan Martinez-Alier's "environmentalism of the poor,"[62] which connects "the poor"—Indigenous and working class—to ecological struggle. Socialist modernism, we can say, is eco-modernism with egalitarian intentions. Huber's variety does not depart from representative democracy, instead advocating the strengthening of electoral political strategies and union organizing.[63] Apparently, degrowth proposals, from this position, are understood as a "hard sell" to the working class and political campaigns, because challenging economic growth and the consumerist lifestyles—or "imperial modes of living"—have become habitual and questioning this is not a popular position in the voting polls. While both agree on some form of democracy, socialist modernism confronts degrowth by asking how their proposed socioecological transition will be accomplished.

On the other hand, the "harder" version of green authoritarianism rebrands dictatorship. The historical conditions of the 1917 Bolshevik Revolution, with all the post-war crisis and turbulence that entails, Andreas Malm equates with the climate emergency of the present to refashion Leninist ideas. Climate catastrophe, said differently, is an opportunity for ecological Leninism to come to power and, despite expected hardships, save the day through an ecologically-oriented dictatorship. This includes implementing, akin to Lenin, an "ecological war communism" that entails:

> learning to live without fossil fuels in no time, breaking the resistance of dominant classes, transforming the economy for the duration, refusing to give up even if all the worst-case scenarios come true, rising out of the ruins with the force and the compromises required, organising the transitional period of restoration, staying with the dilemma. It does not mean cosplay re-enactments of the Russian Civil War. That war deposited a poison

> of brutalised power in the heart of the workers' state, to which it eventually fell victim. Another legacy of the period, however, fared better.[64]

This exposition undoubtedly reads romantic and ambiguous, and any (anti-authoritarian) historian would contest the civil war as the reason for the onset of Leninist dictatorship.[65] The "effective establishment of a one-party dictatorship aggravated tensions between state and society under circumstances of economic and political crisis," explains James Ryan.

> Bolshevik dictatorial rule and the suppression of strikes, uprisings and other socialist parties were not consequences of the White threat [or Tsarist forces[66]]—though this threat certainly helped Bolshevik leaders to justify these measures—and they continued and in some respects intensified after the White challenge.[67]

Regardless, Malm's idea is to seize state power presumably through a coup d'état (like Lenin) or revolutionary movement and to defeat "fossil capital" by aggressively rolling out wind, solar, dams and other energy technologies—which might include nuclear and geoengineering—while creating climate-friendly laws, such as mandatory veganism.[68] This includes deploying the term "fossil capital," that like climate change, reduces the complexity of the problem, creating a single hydrocarbon industry foe and ignoring other capitalist industries and divers of ecosystem degradation.

While confronting and utilizing the state is an understandable desire for any socioecological transformation, preaching an eco-Leninist dictatorship and war communism is alarming to say the least. Noam Chomsky reminds us: "The Leninist system was one of the greatest blows to socialism in the 20th century, second only to fascism."[69] The suppression (even among the Bolsheviks) and the elimination (e.g. execution, imprisonment, and torture) of left revolutionaries and anarchists is appalling—killing and imprisoning all who would resist Lenin's order beginning in April 1918.[70] During Lenin's political rule, between December 1917 and February 1922, conservative estimates suggest "28,000 executions (excluding battlefield deaths) on average per year directly attributed to the Soviet state, a sharp contrast with the approximate total figure of 14,000 executed by the Russian Tsarist regime between 1866 and 1917."[71] This does not seek to make the "West" seem innocent—as the crimes of Euro-America are horrendous—but we do not have so-called critical scholars preaching ecological Churchillism or eco-Rooseveltism. *Lenin's Terror*[72] and *Blood Stained*[73] offer detailed historical exposes, primary sources, and analysis in need of serious consideration before promoting statist dictatorship—left or right. The "State idea, State Socialism, in all its manifestations (economic, political, social, educational)," explains Emma Goldman in 1924, "is entirely and hopelessly bankrupt. Never before in all history has authority, government, or the State, proved so inherently static, reactionary, and even counter-revolutionary in effect. In short, the very antithesis of revolution."[74] Goldman speaks from within Russia at this time, and watched the free assembly, worker self-organization and direct action that was the platform for Lenin's ascendance to power get murdered and disciplined—like the forests, rivers, and mountains—into Soviet state capitalism and its vision of modernization by the Bolsheviks. Said kindly, the cynical carelessness and disregard for the victims of Lenin demonstrated by Malm is antithetical to climate (or ecological) movement building. This promotion of eco-dictatorship by Malm, unsurprisingly, is reinforced by ignoring and slandering political actors—notably anti-authoritarians—and struggles already fighting with their lives, but—of course—would challenge his Leninist dictatorial agenda.[75] Eco-modernism, whether left or right, clings to technological innovation and command-and-control structures, bureaucracy and the resulting socioecological sacrifice necessary to maintain them. The Leninist version, however, pitches itself as 'revolutionary' and remains in a disastrous position.

CONTROLLING SOCIO-POLITICAL TERRAIN: SOCIAL WAR, COUNTERINSURGENCY, AND INFRASTRUCTURAL COLONIZATION

Climate reductionism today is reminiscent of Marx's economic reductionism criticized by anarchists. Marx, we must remember, was also anti-state. The difference, however, was Marx thought, "once classes had been abolished the state would die a natural death, as if through lack of nourishment."[76] Anarchists, on the contrary, believed the state had to be consciously and actively dismantled. Anarchists, moreover, never believed in Marx's "creation of the people's state," "the proletariat raised to the level of the ruling class," and, consequently, the "dictatorship of the proletariat."[77] Seizing the state apparatus Bakunin exclaimed would reproduce the "despotism of a ruling minority" of "*former* workers"—no longer the working class—with a "highly despotic government of the masses by a new very small aristocracy of real or pretend scholars."[78] Recounting Bakunin, Lucien Van der Walt explains: "The sincerity of the revolutionaries was not at issue; rather, the very use of the state machine imposed an 'iron logic' that made state managers 'enemies of the people.' Activists do not change the state; the state changes them."[79] "Scholar" specialists, proclaiming to liberate others and to create a new hierarchical government foretold the rise of Leninism and Stalinism,[80] which Chomsky claims, referring to Bakunin, is "one of the few predictions in the social sciences that actually came true."[81] This speaks to the concerns of Fredy Perlman, agreeing with Bakunin[82] and Foucault,[83] that "the actual proletariat has been as racist as the bosses and the police," observing how oppressed people internalize and pre-project their oppression. Perlman's concluding question—which is also an epigraph for this book—should cause caution against identity essentialism and psycho-political naiveté when they ask: "What concentration camp manager, national executioner or torturer is not a descendant of oppressed people?"[84] Oppression—political, sexual, racial and so on—will not guarantee political commitments toward liberation, let alone the total liberation that struggles against the subjugation and execution of trees, rivers, animals, and mountains.

Marx's privileging of economic factors, or economic determinism, aided in preserving divisions of labor and the refashioning of the state. Marx, Bakunin complained, "takes no account of other factors in history, such as the ever-present reaction of political, juridical, and religious institutions on the economic situations."[85] Bakunin continues:

> Taking into account only the economic questions, he [Marx] insists that only the most advanced countries, those in which capitalist production has attained greatest development, are the most capable of making social revolution. These civilized countries, to the exclusion of all others, are the only ones destined to initiate and carry through this revolution.[86]

Bakunin was refuting Marx's limited (and Eurocentric) theory of development and, as a response, stresses the importance of "the spirit of revolt" that had no borders and can "find it in different degrees in every living being."[87] Marx, replying to this criticism, exclaims "Schoolboy stupidity! A radical social revolution depends on certain definite historical conditions of economic development as its precondition."[88] Black,[89] autonomous,[90] and Situationist[91] interpretations and stretching of Marx's work remain timely and retain synergy with numerous anarchism(s) and autonomous tensions. Anarchism recognizes the necessity for rejecting the "top-down" machinations of the state apparatus, which extends to understanding the struggle against domination is more than just a "class war" that shapes historical development, but a *social war* that identifies the socio-cultural and political techniques that maintain the state and its political economy.

Exploring the birth of the prison, and consequently European state formation, Michel Foucault reveals how "politics is a continuation of war by other means" by inverting the words of military theorist Carl von Clausewitz. "But it must not be forgotten that 'politics' has been conceived as a continuation if not exactly and directly of war, at least of the military model as a fundamental means of preventing civil disorder," explains Foucault in *Discipline & Punish*. "Politics, as a technique of internal peace and order, sought to implement the mechanism of the perfect army, of the disciplined mass, of the docile, useful troop, of the regiment in camp and in the field, on manoeuvres and on exercises."[92] Elucidating Max Stiner's idea of "Prison Society,"[93] Foucault explores the social engineering of European societies by examining how bodies were disciplined, manipulated, and controlled to organize state and capitalist power. The so-called "workhouses" remained instrumental sites for police to take free people—travelers, the so-called "homeless" and "vagrants"—to be tortured into working and learning the dominate moral values of the day—the working class is built with torture and disciplining bourgeoisie morality into people via the police, prison and welfare system.[94] "Organizing and manipulating conditions of hunger," Hannah Kass reminds us, is "an essential weapon of social warfare waged by the state."[95] Land enclosures (or privatization), prisons, and the police emerge as potent instruments of population control and capitalist development.

The seventeenth century Diggers, Foucault tells us, were the first to express social war discourse viewing "the State apparatuses, the law, and the power structures not only do not defend us against our enemies; they are the instruments our enemies are using to pursue and subjugate us."[96] Marxian conceptions of the "ruling class" and "Bourgeoisie state" echoed social war discourse, but the "bourgeois state" framing allowed the state apparatus itself to be salvaged and, thus, ignored the psycho-social processes that reinforce divisions of labor, bureaucracy, operational convenience and power that would attract political leaders and consume revolutionaries. Transforming the oppressed of yesterday, thinking of Perlman, into the torturers of today and administrators of tomorrow. Joseph Déjacque along with André Léo employed the discourse of social war. An anti-authoritarian feminist, survivor of the 1871 Paris Commune insurrection and victim of Marx's First International Purge, André Léo, Gelderloos reminds us, was also a theorist of social war, contending that "the statist, capitalist order constituted a constant war of the dominant classes against all of society."[97] Social war theory would become—often implicitly—developed and practiced by insurrectionary anarchists, articulating the development of permanent conflict against domination. Permanent conflict, implied lived practice of direct action against the state and capital. This, however, included an ardent and highly valuable critique of political organizations, contending that established—and bureaucratizing—political organizations become static in relation to political struggle, prioritizing organizational self-preservation, growth of the political organization itself—emphasizing the number of members over a quality of committed political action—and would eventually betray the struggles and/or issues they were claiming to fight.[98] Political organizations, then, position themselves within the state machinery and are frequently directed by funders, and shareholders and political pressures enable a pacifying and counter-revolutionary function. Any large-scale environmental or conservation non-governmental organization (NGO) could potentially animate this criticism.[99] That is why anarchist, Carlos López "Chivo," reminds us "that which stagnates rots," whether political group or organization.[100]

This critique of political organization, likewise, suggests organizing through affinity groups, which ideally members fluctuate and change, which would create informal networks and/or confederate forms of autonomous networks. This anarchist proposal seeks to find non-dominating forms of political organization to smash capitalism, organize housing, and food provision and, overall, nurture liberated spaces, and the politics of attack. In terms of political organizing, the goal is to remain evasive and invisible to make any attempts at political control difficult, and prolong political struggle. Hence, the state—and its affiliates and surveillance technologies—seek to limit this possibility. The insurrectionary tendency remains an experimental proposal in constant dialogue[101] for articulating a life in permanent conflict. Social war, then, recognizes the low-intensity war by the state against the social relationships of a target population to maintain and extract resources and affirm its organizational existence. This "war" is expressed through dominant institutions, such as schools, factories, prisons, and dominant regimes of city planning designed to crush nature—human and nonhuman—into the machinations of the state and its political economy. The enactment of social war against a people, moreover, remains profound and infects the socio-cultural habits of people and expresses itself in the logic of control itself.[102] In fact, Gelderloos tells us, "the very dichotomy of humans versus nature is a tool, a weapon in that [statist] social war."[103]

Social war theory, it must be recognized, is a theory of colonization. Foucault, exploring the genealogy of "politics by other means" implicitly relates social war to tactics of inter-European colonization.[104] The term "social war," originates from the Roman Social War (91–89 BC), where the Roman Republic learned the benefits of political concessions for maintaining internal stability, as opposed to exclusionary conventional warfare techniques.[105]

Exploring the psycho-social colonization, Joseph Gardenyes discusses how peoples' sense of self (or subjectivities) are altered and that people struggle "against a process of colonization begun in a first instant by ourselves in the form of autochthonous patriarchies and later carried out by a new State and its nascent capital[ist]" institutions.[106] Gardenyes defines social war as "a struggle against the structures of power that colonize us and train us to view the world from the perspective of the needs of power itself, through the metaphysical lens of domination, in which the universe has a center and follows laws and can be quantified and assigned value."[107] The psycho-social and cultural interventions by the state are generational, intimate (as Foucault shows) and continuous. Social war alters how we relate to each other, view ourselves, identify with political systems, and conceptualize political conflict and nonhuman natures. This process of colonization and population control is intimately related to private-public counterinsurgency initiatives and campaigns.

Emerging from the history of colonial warfare, counterinsurgency—Lieutenant Colonel David Kilcullen explains—is "a competition with the insurgent for the right and ability to win the hearts, minds and acquiescence of the population."[108] Here, winning "hearts" is understood as "persuading people their best interests are served by your success," and winning "minds" means "convincing them that you can protect them, and that resisting you is pointless."[109] Counterinsurgency is a type of war—"low-intensity" or "asymmetrical" combat—and style of warfare that emphasizes intelligence networks, psychological operations (PSYOPS), media manipulation, and, finally, security provision and social development that seek to maintain governmental legitimacy.[110] Insurgency, defined by the *Insurgencies and Counterinsurgencies Field Manual*, is when "a population or groups in a population are willing to fight to change the conditions to their favor, using both violent and nonviolent means to affect a change in the prevailing authority, they often initiate an insurgency."[111] Nonviolent protest, then, is understood as a threat to government and political order. Kristian Williams makes the distinction between "hard" (direct) and "soft" ('indirect) practices of counterinsurgency that work together in larger strategies of population control.[112] Hard techniques—the proverbial "stick"—include overt political violence by police, military and mercenary forces, while "soft" techniques—the "carrot"—are civil-military operations that invest resources and technologies into underdeveloped or "troubled" areas." Soft" interventions are commonly referred to as civilian assistance, community and social development, including foreign aid provided to collaborating local elites to stabilize and manage areas of interest. Social war, and counterinsurgency, are central processes by which internal and external colonization is executed. This is why Kilcullen reminds us that "counterinsurgency, then, or counter-insurrection, seems to be an enduring human social institution that has been part of the role of virtually every government in history and perhaps even partly defines what we mean by the word state."[113] The application of counterinsurgency techniques is ubiquitous within the military, police departments, and mercenary organizations, which is training that transfers into narcotics trafficking organizations (aka Narcos) and other extra-judicial groups. Marketing agencies are also inspired by counterinsurgency doctrine,[114] and vice versa in matters of psychological warfare and propaganda campaigns. These groups, moreover, retain various institutional allies and supports, depending on the political context and their objectives.

By approaching wind turbines, mines and energy infrastructure from the perspective of social war, this book is rooted in the perspective that political repression, racism and ecological degradation are avoidable. That is socioecological catastrophe *is not enviable, can be stopped* and *socio-political systems do not have to be this way*. Call it idealist, but this perspective allows us to look at mining for what it is—and doing—critically evaluating the claims of particular projects; the desires they manufacture or nurture; the opposition that emerges; the results of the money "invested" into communities; and the general outcome, even if development projects are always in process. Insurrectionary research approaches, it must be recognized, might bring us closer to this mythical objectivity by critically interrogating the sacred concepts of the "State," "technological progress," and "modernity" to really explore its "dark side" or, in business terms, the "cost" of statism, modernism and techno-capitalism on people and the planet. Given the rate of socioecological catastrophe, we must critically assess modernist infrastructures, modes of habitation and development of people and, certainly, researchers from well-to-do universities. The lens of infrastructural colonization, which could be called the material enactment of social war, explores how territories are controlled, developed and transformed by megaproject infrastructures. While colonization might be an inflammatory term, it describes the socio-technical regime and modality consistent from European state formation to the classical colonial and (neo)colonial present. The colonial is understood as a model of spatial organization, environmental relationship, racial discourse, and political program that has only developed and evolved with (statist) governance and (capitalist) innovation. Before colonial expeditions and conquest could be executed overseas, similar processes of warfare, territorialization and politico-economic integration had to be performed at "home" before colonial powers could emerge. This is why statism is understood as colonialism, projecting similar political, economic and military structures across the world.

Modernist infrastructure remains an essential part of this structure of colonial conquest—at "home" or "abroad." The concept of infrastructural colonization explores social warfare, dissecting the ideology of modernity, discourses of progress and the fabrication of desires/aspirations of populations—near and far—to justify its socioecological extraction and so-called development. This lens immediately confronts the naturalized self-identification with statist and capitalist systems, which typically ignore socioecological oppression in favor of capitalist normalcy, its enchanting entertainment and convenient technologies. Colonial and capitalist systems, as much as they employ repressive and torturous technologies, also fabricate enchanting dreams, desires and technologies. Social warfare theory, thereby, recognizes that invasion is infrastructural and will structurally alter the ecological, social, cultural and economic compositions of a particular territory. *This System is Killing Us* remains an exploration into social war and the enactment of infrastructural colonization, because the common feature that connects these chapters together is that people viewed these projects as attacks, encroachments and colonizing forces that necessitated resistance.

GETTING INTO THE BOOK, THE METHODOLOGICAL APPROACH

The research for this book began a decade ago, in late 2013. This work began to focus on wind energy development in Oaxaca, Mexico, as part of my doctoral dissertation and, later, other articles, which are revisited in Chapter 2. *This System is Killing Us* is an overview of my research, threading a narrative through various case studies that have been my political, but also life journey through numerous environmental conflicts. The themes from the case studies have been discussed previously in academic journals, and some popular pieces, but—overall—the findings from what has been intensive, life-threatening and engaged fieldwork have been largely ignored within the current popular academic debates discussed above, not to mention the public at large. This book serves to join these conversations, revisiting these sites and threading this work together to bring forth another perspective to the conversations on climate action and socioecological transformation.

This book relies on participant observation, ethnography, and various types of interviewing. Participant observation is about being in places, sitting, listening, experiencing, watching, and participating in the sites of research. These methods, however, are not as simple as just hanging around; they entail attentively taking field notes, journaling about the day, writing meta notes to hypothesize different theories and political-personal connections, and, later re-reading, reflecting, and thematically categorizing the emergent themes and issues from interviews and daily life in areas of research. These reflections, moreover, were done in conversation with numerous people from these areas and elsewhere. This methodological approach departs from "armchair" theorizing, data points, and modelling constructions. Ethnography, completely complementary to participant observation, is described as "deep hanging out" or literally means "writing about people."[115] Describing the (ideal) ethos behind participant observation and ethnography as they coalesce into anthropological methods, Tim Ingold explains:

> For to observe is not to objectify; it is to attend to persons and things, to learn from them, and to follow in precept and practice. Indeed there can be no observation without participation— that is, without an intimate coupling, in perception and action, of observer and observed. Thus, participant observation is absolutely not an undercover technique for gathering intelligence on people, on the pretext of learning from them. It is rather a fulfilment, in both letter and deed, of what we owe to the world for our development and formation.

As Ingold alludes with "an undercover technique for gathering intelligence on people," anthropological methods are intentionally misused to support the military, police, marketing agencies and, as we will see throughout the chapters, mining companies. Even if many are sociologists (and not anthropologists)—extractive companies rely on social scientists and agronomists to construct strategies to subdue protests and manufacture relative consent to permit extractive development. It is for these continued, and past-colonial histories, I have taken a position of anti-anthropology. Anti-anthropology challenges the "foundational 'home-field' dichotomy that tends to deny the connection between where the researcher originates and the area where research is conducted," but also extends this self-critical gaze to what the university system and what anthropological knowledge produces.[116] Ingold's understanding of anthropology is the ideal, while anti-anthropology does not challenge this ideal, it does challenge its institutional collaborations past and present with extractive industries and governments. Anti-anthropology, contrary to colonial or mainstream anthropology, seeks to organize knowledge not for the academy, but for struggles and individuals taking a position of permanent conflict near and far in defense of enriching cultural practices, social fabrics and ecosystems. This book seeks to maintain generational lessons, case study examples and general insights into the complications and deceitful realities of political conflict and survival.

In response, especially in politically volatile environmental conflicts, this means researching the companies and *not the people.* This attempts to flip anthropology's colonial legacy on its head, by studying the perceived invaders and not the people remains enduring in the study. This tension, however, has existed within anthropology at least since the 1970s with Laura Nader's idea of "studying up" the ladder of power.[117] The people in resistance, and their environments, often fighting multi-million or billion-dollar companies or projects, are fighting a difficult and uphill battle. Research for this book maintained the intention to investigate the companies, and understand the conflicts, perspectives and tensions from people to learn why they are fighting, negotiating or celebrating megaprojects. This, overall, *means not researching the resistance per se, but the projects and companies themselves.*[118] Obviously, overlap emerges, yet the anthropologists or political ecologists muddling in the affairs of Indigenous groups or other land defenders is generally not helpful to them—I have learned this the hard way. This translates into only writing well-known aspects about the resistance or providing details worth knowing about events, often to show the determination and life-threatening attempts of people to stop the invasion of police, politicians, and the infrastructure projects or mines they are promoting in these areas. This research, it turns out, is about revealing the violence experienced by land defenders, their determination and the multi-dilemmas and internal conflicts they confront.

While this book is single-authored, the research in all of these areas was conducted with friends and people who invited me into these areas and believed in my research projects. In Oaxaca (Chapter 2), there was nearly an entire collective, most of which remain purposely nameless and evasive. In Germany (Chapter 3), Andrea Brock was the lead researcher on the Hambach forest lignite coal mine, but there were also supportive friends living in struggle in the forest. In Peru (Chapter 4), it was Carlo Eduardo Fernandez who invited and worked with me in the Valle de Tambo region of southwest, Peru, to investigate the Grupo Mexico mining subsidiary Southern Copper Peru's highly contested Tía Maria mine. In France and southeast Iberia (Chapter 5), it was Jean-Baptiste Vidalou, Louis Laratte, and numerous other individuals assisting and supporting this research with their time, energy, and contacts—with more remaining anonymous. Finally, in Portugal (Chapter 6), Mariana Riquito conducted research with me in this area and, later, would continue their doctoral fieldwork in the

Barroso region. Together, we would employ 'snowball' sampling, which means we had friends in the region to interview and from those interviews would get more contacts, the process had the effect of accumulating interviews like a "snowball" rolling down a steep hill collecting snow. This also includes walking around going door-to-door haphazardly to interview people or approaching people in public spaces. This, moreover, was complemented by strategically contacting government officials, company representatives, and NGOs to interview them across all the sites. This also included walking, riding bikes, and being driven around sites of energy or mineral extraction. This was a hyper-proactive approach that included conducting informal interviews while hitchhiking, riding on public transportation and at public events.

All the research projects employed participant observation or, more accurately, a type of participating observation. This entailed, in many instances (Mexico, Germany, France), being with and in the resistance to understand the situations they were facing, which often entailed various intensities of intimidation and violence either by police, mercenaries, or company representatives. Research accompanied immersion into different cultural events, parties and the like. Participant observation was complemented by interviews, specifically semi-structured, informal and oral history interviews. Semi-structured interviews had a general set of questions and focus, but remained flexible and depended on unanticipated follow-up questions to support unanticipated tangents or concerns. Informal interviews, likewise, follow that general script but are usually done within informal conversation and written into field notes afterwards. Oral history interviews, while maybe the least common within the book, were interviews conducted with people whom I often interviewed more than once, and would follow them up by interviewing them about their lives to better understand their personal struggles and how they landed where they are in life. The number of informal interviews behind this book is almost incalculable,[119] but could easily be over 150 across all the chapters. The semi-structured interviews conducted in Oaxaca, Mexico, equaled 164 between the years 2014 and 2020. In Germany 15 semi-structured interviews were collected by Andrea Brock (who continued this work for the following years); 47 interviews were collected in Peru, while 74 interviews were collected in France between 2018 and 2020; 26 in Catalonia and Spain; and 28 interviews in Portugal. Additionally, Andrea Brock was interviewed for Chapter 3 for an updated understanding of the anti-coal mining movement in Germany. Likewise, there have been countless discussions with NGO representatives and research modelers, which corresponds to the next chapter critically assessing the knowledge production around socioecological crisis, to which we now turn.

1

The science of maintaining socioecological catastrophe

Public awareness, and general knowledge, of climate change circulates more frequently in quantity and quality than ever before. This, in part, is due to the increased saturation of information technologies into personal, but also public, spaces—from smartphones and social media to televisions in bars, trains, businesses, and increasingly elaborate advertisements in city centers and anywhere space and local regulations permit. Why might greater awareness, concern, and talk about climate change fail to address this planetary issue? Some might shoot back: "It is a slow process"; "We are on the right track now with green technologies"; or "Change is happening as we speak"—whatever "change" actually means. The ecological catastrophe is publicly acknowledged—no thanks to hydrocarbon industries retarding the process with public relations propaganda and lobbying[1] that promotes climate denialism. This propaganda confuses people, providing the excuse they want to believe in: capitalist consumerism is "not that bad" and the government will "fix it." Meanwhile, socioecological catastrophe persists as international carbon benchmarks are repeatedly failing[2] (see Figure 1.1). It seems there is a correlation between "raising awareness" within the mainstream channels of modernity and the prolonging of climate catastrophe.

This book shows how socioecological catastrophe continues despite this public awareness, media time, scientific expenditures, and public "lip service" dedicated to these issues. This chapter, contrary to the rest of the book, will examine numerous factors at the cultural and systemic level that lay the foundation for this socioecological failure. This occurs at three levels. First, the cultural myths propagated by governments and industry to maintain and cement a particular political economy in place. This cultural myth, second, is reinforced by ontological assumptions and beliefs—or views systematically indoctrinated into people to ignore and abuse the life forms around us that sustain human populations. And, third, this ontological worldview enables the construction of a particular epistemology—the method of knowledge generation—and scientific tools that enable conservative estimates, partial accounting and, overall, place ecosystems and their inhabitants at a disadvantage in favor of capitalist political economy and statist governance.

Given the high theory related to the philosophy of science, I want to be clear about the meaning of ontology and epistemology. Ontology is the study of being, a branch of metaphysics concerned with the nature and relations of existence.[3] Ontology, then, is the study of the "real" and what can be said to exist. Epistemology, on the other hand, is the study or theory of knowledge creation that explores the origin, methods, and limits of human knowledge. Foucault's famous book, *The Order of Things*,[4] challenged modern epistemology and coined the term *episteme* to reveal how scientific method, knowledge, and what constitutes "Truth" is socially constructed and, often, situated with a particular ideology. This is why Christopher Newfield and colleagues affirm: "It is now uncontroversial that what, how, and why we count is always shaped by our values and interests."[5] Claims to "objectivity" and "truth"—especially in the realm of scientific method—are actually subjective and the social construction of "a truth." Objectivity is impossible and, at best, the study of anthropology recognizes that the best one can do is be as detailed as possible about their methodological approach, implicit assumptions—technical, political, and ideological—and offer background on the researcher and their intentions (e.g., positionality). Even this, as discussed below, often leaves political, ideological, and developmental bias intact.

While this chapter focuses on environmental accounting, it must be stressed that this is not separate from social realities. Ecological problems are deeply embedded with social issues—racism, hierarchies, and discrimination—and how we treat each other is deeply related to ecological issues. As has been shown elsewhere, nearly all struggles against injustice are at the same time ecological struggles. Racial discrimination and abuse by the police, for example, are only possible by subjugating, mining, and processing ecosystems to create weapons and institutions to impose inequality and violence.[6] This alludes to structural issues embedded in, what Carolyn Merchant[7] and Vandana Shiva[8] call, "modern mechanical science" and how it approaches people and environments. Focusing on scientific methods, this chapter encompasses a focus on the "social" within "socioecological" that is responsible for aiding and abetting the maintenance and continuation of ecological catastrophe.

THE CULTURAL MYTH OF SUSTAINABLE DEVELOPMENT AND GREEN GROWTH

Ecological devastation from industrial development began to sink into the minds of people and policymakers in the 1960s. Native American Traditions, global Indigenous struggles (in their great variety), and Rachel Carson's *Silent Spring*[9] were seminal in spurring environmental consciousness, leading to the emergence of an environmental movement—environmentalism—in the Global North and some of the most comprehensive environmental legislation in the late 1960s and early 1970s in the US and Europe. This included Edward Nixon persuading his brother, US President Richard Nixon, to create the Environmental Protection Agency (EPA) and, more interestingly in the 1980s, creating joint ventures with Chinese companies to produce rare earth mineral-based permanent magnets, instrumental for cars, computers, and wind turbines. This entailed offshoring the toxic reality of rare earth mining, processing, and manufacturing to China,[10] which today dominates the heavy rare earth element market—with Europe 97% dependent on Chinese industry and mining in 2020.[11]

In 1972, the Club of Rome Report, *The Limits to Growth*,[12] formally recognized the need to degrow energy consumption and material production, as well as stabilizing global populations. Population growth has been used to advance racist discourse and programs targeting Indigenous and Black populations for sterilization, meanwhile conveniently ignoring population growth and consumption patterns among affluent demographics.[13] As the one frequently cited statistic, from Our World in Data, contends: "1 person in the US consumes as much energy as 2 French, 11 Indians, 24 Ghanaians, 82 Afghans, and 329 Somalis."[14] While there is likely, much to dissect in this statistic, material extraction, manufacturing, and energy use stresses the world more than population *per se*—even if the two are related. Despite concerning shortcomings, the *Limits to Growth* report represents peak acknowledgment of socioecological devastation emanating from global capitalism creating an important foundation for policy changes. That is, until the concept and program of sustainable development arrived in the 1980s.

Sustainable development, and its offspring concepts—such as "green growth," green economy, the "Scandinavian Model"—remain violent statist capitalist myths that have delayed and prevented the possibility of adequate environmental policy to address ecological and climate catastrophe. Sustainable and green economic projects are often pitched as "win-win" scenarios that mitigate ecological degradation and generate income, if not profits, from conservation, tourism or energy development schemes.[15] While most people—myself included, until ten years ago—want to believe in sustainable development, it is nothing more than a method to renew capitalism and continue the existing planet-threatening trajectory. While the term was percolating in the early 1980s, it was not until 1987 that it made its formal debut in the United Nations *Our Common Future: The World Commission on Environment and Development* or Brundtland Commission—headed by Gro Harlem Brundtland, Norway's first female prime minister (years 1981, 1986–1989 and 1990–1996).[16] Sustainable development, the report defined as meeting "the needs and aspirations of the present without compromising the ability to meet those of the future." The Brundtland Commission continues, "Far from requiring the cessation of economic growth it recognises that the problems of poverty and underdevelopment cannot be solved unless we have a new era of growth in which developing countries play a large role and reap large benefits."[17] Mixed with the good intention of meeting "the needs and aspirations of the present without compromising the ability to meet those of the future," sustainable development is conceptualized as "a new era of [economic] growth" that will include "developing countries" in on the spoils of extractive development. Poverty alleviation is used to ignore how poverty is measured and sustained to position capitalist growth as the answer to poverty and environmental issues. Meanwhile, the "rich," capital accumulation, and the present modality of development are not identified as sources of socioecological crisis. "Twenty years ago some spoke of the limits of growth," exclaims US President George Bush at the 1992 Rio Earth Summit, "*And today we realize that growth is the engine of change and friend of the environment.*"[18] Despite the rhetoric of the environment, poverty reduction, and so forth, the focus is on preserving capitalism in its present articulation, creating new markets and—most of all—implicitly refusing critical self-reflection on extractive development, market relations, and the harms of modernist infrastructure and governance. Sustainable development and the rhetoric that "growth is the engine of change and friend of the environment" has remained a potent mechanism to continue the present socioecological trajectory by other means—or really by placing green labels on capitalist bottles of wine.

Sustainable development would give way to ideas of Payment for Ecosystem (PES) that sought to further expand the grid of private property over ecosystems by "selling nature to save it."[19] The idea was to expand private property, ecotourism, and get people to invest in conservation to "save the planet." While consistent with capitalist logics of ownership, exclusion, and profiteering, it has made little sense to ecosystems and people closest to the land. Market-based conservation has continued colonial, and in many instances genocidal and ecocidal, trajectories that seek to control land, employing a combination of dispossessing people from their forests or integrating them into conservation or small business schemes. Market-based conservation has generated increasing conflict, human rights violations—among them systematic rape of inhabitants—and the variegated failure and limited successes of these programs.[20] Sustainable development, importantly here, is legitimized by the idea of "decoupling."

Decoupling, contends, that you can separate—or decouple—ecological degradation from economic growth. *Relative decoupling* is when economic growth is faster than the growth of environmental

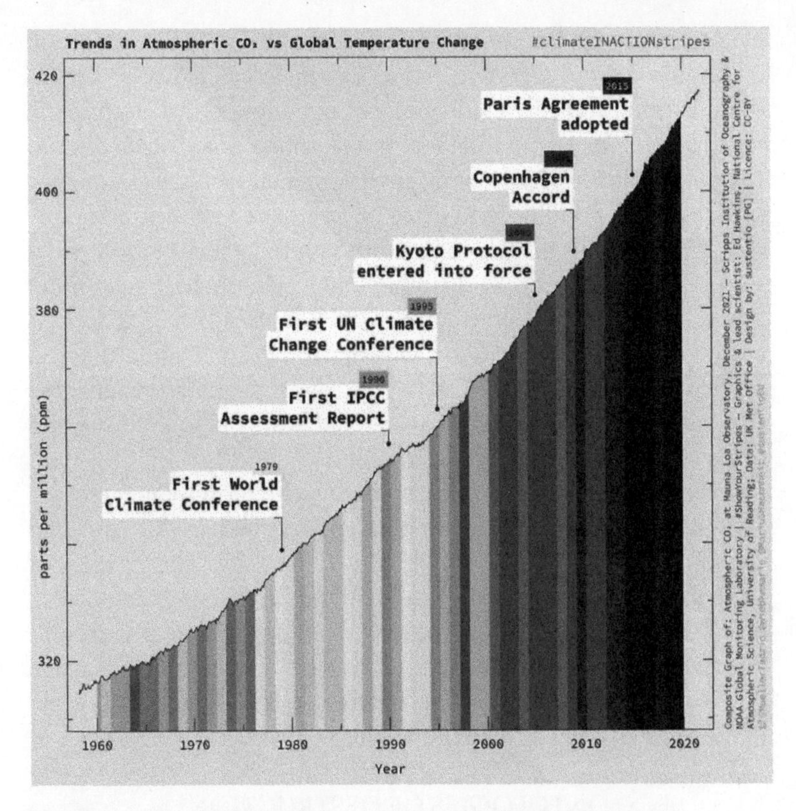

Figure 1.1 Trends in atmospheric CO_2 versus global temperature change, 1958–2020, with climate conference dates indicated. Source: #climateINACTIONstripes Graphic: @MuellerTadzio / @wiebkemarie / @MariusHasenheit / @sustentioEU

damage, but ecological degradation continues at a steady pace. *Absolute decoupling* is the ideal, when the economy grows while the amount of resource-use and/or environmental impact is decreasing.[21] While political ecologists have been systematically dissecting and refuting these claims, investigating conservation, ecotourism schemes, and low-carbon infrastructure since the early 2000s, belief in sustainable development and, later, green growth would continue to dominate, which—as mentioned in the introduction—has now systematically been discredited by ecological economists and degrowth scholars. While the introduction, drawing on degrowth scholars, explains how there is extremely weak or no evidence to support assertions of decoupling, Tere Vadén and colleagues remind everyone that "the empirical evidence on decoupling is thin" and "the evidence does not suggest that decoupling towards ecological sustainability is happening at a global (or even regional) scale."[22] The growing environmental conflicts, frequency of forest fires, flooding, and extreme weather events are testimony enough. Not to forget the continuous failure to curb carbon emissions (see Figure 1.1). Said simply, the way industrial economies are extracting from people and ecologies is already having serious socioecological repercussions, which green growth and ecological modernist discourses attempt to conceal or minimize by ignoring this reality or asserting it can be fixed with nuclear power and more of the existing energy technologies.

Overall, people cannot equate mining, chemical manufacturing and the widespread toxification of environments with so-called "sustainability." To put it simply, this is insanity. The realities of concrete buildings, asphalt, producing electronics, vehicles, and, even, wind turbines are significantly degrading, as we will explore in greater detail in later chapters. This normalization of extractive development, and its negative outcomes, relates to how people make sense of and relate to the world.

ONTOLOGIES OF DESTRUCTION

Modern science imposes a particular ontology and materialist epistemology onto the world. Explicit about its desire to subdue, dissect, and control nature,[23] modern science seeks to create universal truths and dominate the epistemological and, consequently, the ontological realm.[24] Central to modern mechanical science is the principle of separation. The "Cartesian dualism," which separates mind from body—viewing them as ontologically distinct, yet interacting—remains central to Western thought and science. This principle of separation, Robert Romanyshyn[25] argues, begins in art history when in 1425 Italian painter Filippo Brunelleschi invented, or better formalized, "linear perspective vision." Linear vision separates the painter from landscapes—decontextualizing and creating "the spectator behind the window"—to apply geometric standards (e.g., vanishing points, distancing points, horizontal, and vertical line alignment) to create a grid technique to promote accurate painting scales.[26] "The geometrical, the quantifiable, the measurable dimensions of the world become primary" within a linear perspective, explains Romanyshyn, which, furthermore, necessitates "reduction, of miniaturization, qualitative dimensions are destined to become only secondary."[27] This artistic psycho-political technology was the precursor to the anatomical, medical and, arguably, biopolitical gaze underlining modern science.[28] It should be recognized that this technology underlines mechanical science, creating great possibilities, but it also came to dominate, subordinating, if not eliminating other sciences and knowledges, and became associated with empire.[29] Outlining Descartes's methodological precepts that root Western epistemology (e.g., science), Carolyn Merchant (1983: 231) summarizes:

1. To accept as true only what was so clearly and distinctly presented that there was no occasion to doubt it;
2. To divide every problem into as many parts as needed to resolve it;
3. To begin with, objects simple and easy to understand and to rise by degrees to the most complex (abstraction and context independence);
4. To make so general and complete a review that nothing is omitted.

Materialism, separation, division, and dissection become important epistemological tools. This separation of mind/body, humans/nature, or culture/nature are foundational to modern science, which Mario Blaser explains, if "moderns" have "the confidence that they have more than a perspective" and recognizing such dualisms "allowed moderns to develop the proper procedure for knowing reality as it is: universal science."[30] The standards and criteria that justify modern science, and its universality, are socially constructed by particular parameters based on a particular ontology and, consequently, cultural values and moralism. Speaking to thermodynamics, Cara Daggett explains: "The European pursuit of work and waste, and its reliance on racist justifications, predated the science of energy, but energy reinforced them by mapping these virtues onto the efficient operations of fossil-fueled machines."[31] Science, from thermodynamics to biomedicine, was organized and adapted to the prevailing politics, ideologies and the project of colonial development.

Modern science has obviously proven effective, reliable, and created a world of technological wonder. Yet, the question is what have been the consequences of this ontological and epistemological perspective? What have they prevented people from knowing, feeling, and seeing? What knowledges and, corresponding, practices have been buried under the hubris and claim to supremacy of modern science? Said differently, what cultural values and ontologies have come to rule? The standards of validation and criteria for affirming a particular ontology and epistemology become slippery. In order to justify human supremacy (anthropocentrism), Sian Sullivan offers the gruesome example of "life scientists" who "shoehorn" ontology by severing the vocal cords of animals "so that they would not be able to hear their animal cries of pain" and therefore validate the inferior communicative capabilities of nonhumans.[32] Following this, we might see the green economy as an attempt at severing the anxieties and actions of humans and nonhumans resisting and attacking, or potentially, taking action to stop the present trajectory of techno-capitalist development and its ecological and climatic consequences. The mechanics of green capitalism then rely on the constructions of energy, biodiversity and carbon that organize quantitative scientific methods to justify the continued conquest of nature.

EPISTEMOLOGICAL CAVEATS: ENERGY, BIODIVERSITY, AND CARBON

Energy

Where did "energy" come from? Energy emerged as a concept in the 1840s as coal-fed steam engines were multiplying across Europe.[33] The social construction of "energy" emerged as a technology of the Industrial Revolution, which sought to make visible, categorize and measure vital life resources of nonhumans and humans to economize and harness their fuel and power. Energy as a signifier and its science, thermodynamics, as Cara Daggett shows, had to negotiate contradicting and challenging Protestantism—the dominant religion in England.[34] In this way, Daggett shows, thermodynamics was less challenging to Protestantism than evolutionary theory. It is no surprise then, that the fabricators of "energy" were believers in the Protestant work ethic, utilitarianism, government, patriarchy, and racist stereotypes.[35] Energy and physics emerge to reinforce exploitative labor regimes. Labor exploitation predates the category of energy, but in the nineteenth century, "you see a move in the language from something that is of a more religious discourse, about 'civilizing' and 'Christianizing' people, toward a scientific discourse about a way of working on bodies and disciplining energy flows," explains Daggett.[36] Energy emerged as a new source of legitimacy, portraying itself as natural, normal and "the way of the world." People, Daggett explains, "could be categorized according to their energy, their productivity, their efficacy and assumptions can be made about people who needed governance, so they would not be 'wasteful,' 'idle' or 'lazy.'"[37] Categorizing people, but also landscapes, as "wasteful" or "unoccupied" projected a colonial gaze over lifeways, eco-systemic complexity, and culture to discursively erase this life from the land to declare it *terra nullius* ("nobody's land"), then grab it and regiment people and ecosystems to the colonial economy.[38] While thermodynamics is functional, it transfers and embeds a particular value system into a science. Energy, in many ways, became a moral construct and justification for global divisions of labor that "naturalized the imperial circulation of power, which sacrificed people and things to the project of work, just as coal was sacrificed to the engine."[39] Thermodynamics, and energy as a unit of measurement, economized patriarchy, advanced utilitarianism, and the creed of economic growth, which was the enlightened perspective spread across the world through colonial venture. This epistemological logic proved instrumental to colonizing human and nonhuman resources at "home" and "abroad."

Energy became the backbone of techno-capitalist development. "Energy is a thoroughly modern thing," Daggett reminds us.[40] While the distinction between fossil fuels and renewable energy stretches back to the Industrial Revolution,[41] the entry of energy into politics did not formalize until the 1973 oil crisis.[42] Energy as a political field corresponds with the oil crisis, but also the environmental crisis that gained traction in the 1970s. The US Department of Energy formed in 1977, which Daggett explains mainstreamed concepts such as "energy transitions," "energy alternatives," and "energy forecasting" paved the way for "energy companies" and corresponding "energy outlooks" that guide industrial policy today.[43] More critically, however, historians of science, Christophe Bonneuil and Jean-Baptiste Fressoz explain, "If history can teach us one thing, it is that there never has been an energy transition," in fact it is "successive additions of new sources of primary energy."[44] Bonneuil and Fressoz show us that the term "energy transition" was "invented by think-tanks and popularized by power institutions: the US Department of Energy, the Swedish Secretariat for Futures Studies, the Trilateral Commission, the European Community and various industrial lobbies."[45] "'[T]ransition' rather than 'crisis' made the future less generative of anxiety, by attaching it to a planning and managerial rationality," explain Bonneuil and Fressoz.[46] The phrasing of transition ignores the continuation and addition of generalized energy extraction,[47] which is motivated by numerous factors. Among them, preserving capitalism, statist governance, and opening the doorway to environmental-technological optimism and, consequently, market development in line with the rise of sustainable development. Energy, as a concept, remains essential to enabling land grabbing and extractive activities. The colonial logic and program, as we will see in subsequent chapters, continues in full force but now in the name of saving the environment and mitigating climate change. Let the conceptual history of energy serve as a foundation for contextualizing the different infrastructure and extractive projects we will explore throughout the book.

Biodiversity

Biodiversity is a catchall phrase that refers to the lives of various species. Replacing the terms "species diversity" and "species richness,"[48] biodiversity rose to prominence after the UN WCED1992 "Earth Summit" and includes terrestrial, marine or aquatic biodiversity. The discourse of biodiversity, Arturo Escobar contends, transforms nature into "a source of value in itself."[49] The term biodiversity transforms species of flora and fauna into "reservoirs of value that research and knowledge, along with biotechnology, can release for capital and communities."[50] The scientific endeavors that categorize, monitor, and, eventually by the UN, NGOs, and market actors, organize ecosystem valuations do not protect them. It divides and conquers ecosystems. Biodiversity, with Western frameworks, tends toward separating the "Web of Life" into pieces to be monitored, managed and creating hierarchies between "species" or different existences[51]—that is only intensified by monetizing conservation (see below). This separation emerges in conservation by labeling Indigenous subsistence as "poaching," thereby justifying their dispossession and separation from the landscapes, rivers, and trees they have been living with and protecting for centuries.[52] This, ironically, creates greater opportunities for commercial poachers—the ones we think about with helicopters and high-powered rifles—and with the arrival of conservation, weakens territorial defense by locals (as they have been relocated out of the forest), and requires costly militarization of the area (that frequently results in human right violations and abuses).[53] The biodiversity framework, and conservation, creates visibility for management, commodification, and, in many instances, the progressive—"conscientious"—destruction of those ecosystems.

In line with the creed of sustainable development, biodiversity is a categorical priming for the objectification of nature. The "programmes, or international meetings of donors and policymakers to discuss the fate of 'the global environment,'" Sullivan explains:

> thus requires and reproduces acceptable conceptualisations of, and relationships with, the presentable, packageable, consumable and manageable objects of "nature," "biodiversity" or "the environment." A "nature" with which human relationships are reduced to sustainable consumption and custodial practices, whether direct or indirect, for livelihoods or for profit.[54]

The United Nations has been instrumental in this commodification process. The 2005 United Nations Millennium Ecosystem

Assessment report, is one of many examples, which conceptualizes nonhuman natures as "ecosystem services" and offers 24 "service categories," such as provisioning services (food, water, timber, fiber, etc.), regulating services (floods, droughts, land degradation, and disease), and so on.[55] This conceptualization of nature as a service provider, in the words of Sullivan, "begins the discursive and conceptual transformation of earth into a corporation, providing goods and services that can be quantified, priced and traded as commodities."[56] This discursive and conceptual transformation, however, is not innocent.

Conservation sites are militarized, employing counterinsurgency to enforce "park" boundaries once home to Indigenous inhabitants.[57] Conservation in Northern America, and elsewhere, has an exterminating history. The idea of pristine nature—nature devoid of humans—happened to complement the frontier wars and genocidal campaigns against numerous Indigenous nations.[58] And this is the important lesson, people are a part of ecosystems—Indigenous nations were central to cultivating Amazonian, African, and Redwood forests.[59] Humans can play an important and positive role within their ecosystems—if they want to and decide to cultivate beauty with other nonhuman and more-than-human natures. Instead, the marketization of conservation discursively captures, encloses, marketizes, and financializes non-natures in the name of biodiversity and climate change mitigation.[60] "'[B]iodiversity conservation' and other 'natural climate solutions,'" Philippe Le Billon demonstrates, "represent dangerous tools of land reallocation, creating spaces of exception and annihilating 'traditional' socio-environmental forms of life."[61] Le Billon reviews how conservation and extractive industries are advancing multiple forms of extraction, demonstrating how "biodiversity offsets can pave the way to opening up new spaces of coal extraction" as well as "offer the agro-industrial group opportunities to further coerce local communities into limiting their traditional livelihood activities and possibly evict them from their ancestral territories."[62] This also includes how mining for low-carbon infrastructures is increasing the rates and trajectory of mining conventional (e.g., iron ore, copper) and critical raw materials (e.g., rare earth minerals).[63] Said simply, the green economy, through the framing of biodiversity (and carbon), is advancing ecological extraction and political control.

The trend of "selling nature to save it" overlaps with "neoliberal multiculturalism," which Hale describes as redirecting "the abundant political energy of cultural rights activism, rather than directly oppose it."[64] Environmentalism and Indigenous resistance have been undermined by recognizing and including nonhuman and human natures into statist and capitalist political economy—inclusion tends to further divide, conquer and degrade people fighting to resist the roots of ecological conquest: colonial occupation, land privatization, Western onto-epistemological approaches to ecosystems and its integration with market mechanisms. Escobar foreshadows neoliberal multiculturalism and nature with the concept of biodiversity. "[T]he tropical rain forest areas of the world," Escobar explains, "are finally being recognized as owners of their territories (or what is left of them), but only to the extent that they accept to treat it—and themselves—as reservoirs of capital."[65] Forests and habitat are ontologically flattened and reduced to investment "resources," ushering in new waves of socioecological colonization and growing the national economy—not the forests and people of those lands, even if they receive residual benefits (e.g., the "trickle-down effect"). The green economy extends the trajectory of techno-capitalist progress, even though it provides numerous benefits, which makes this a subject of intense debate among scholars.

Carbon

Carbon dioxide remains the principal unit of measurement for evaluating global warming and justifying the green economy. Carbon dioxide, or carbon, calculations are essential to low-carbon infrastructural development, but also conservation, specifically with carbon sequestration schemes like REDD+ (Reduced Emissions for Deforestation and Forest Degradation). Carbon measurements represent seven different greenhouse gases (GHG): water vapor, carbon dioxide, methane, nitrous oxide, hydrofluorocarbons, perfluorocarbons, and sulfur hexafluoride. In 2001, the "Greenhouse Gas Protocol: A Corporate Accounting and Reporting Standard," was designed by the World Resource Institute (WRI) and the World Business Council for Sustainable Development (WBCSD), which currently serves as the main framework for emissions accounting worldwide.[66] The WRI and WBCSD update how they calculate commensurability between carbon and the six other GHGs, for example, methane rates change by 29% over 20 years.[67] Carbon serves as the general metric to chart global warming, attempting to make these gases commensurable with each other to offer approximations of global warming and climate change. Carbon accounting and modeling, however, are subject to, at least publicly, enormous uncertainty and inaccuracies. This retains serious consequences when carbon accounting remains the principal framework for charting the ecological footprints of companies and climate change.

Carbon, like energy and biodiversity, has the same ontological and epistemological underpinnings. Carbon is a colorless gas and element that naturally exists within the atmosphere. There are, according to Tasseda Boukherroub and colleagues, four methodologies for measuring the carbon footprint.[68] First, "direct measurement" that directly measures pollutants from sites, yet typically is only done if regulated by authorities. Second, "energy-based calculations" are "based on mass balance or theoretical combustion specific to a facility or a process." This measurement is applied to fuel consumption. Third, "activity-based calculations" apply an activity-based cost formula that calculates a total cost pool and divides it by the cost driver, which yields the cost driver rate. The cost driver rate is an activity-based counting that calculates the amount of overhead and indirect costs related to a particular manufacturing or extractive activity. Information is frequently withheld, sometimes due to competition between companies or the potentially damaging effect it can have on a product.[69] Fourth, economic input–output lifecycle assessments' (EIO–LCA) models convert company expenditures into the average amount of carbon emissions. "Carbon data" is often provided by companies themselves, or part of online data platforms that are updated and people can subscribe to plug this data into (various) lifecycle assessment models.[70] This, we should note, raises serious concern about data collection and what actually constitutes the data being plugged into these models. Furthermore, the responsibility of companies is difficult to identify, because projects are jointly owned and, according to Yann Bouchery and colleagues, "'companies' direct emissions average only 14% of their supply chain emissions prior to use and disposal."[71] Companies have intricate webs of business subcontracting, which creates significant challenges and potential zones of plausible deniability for the socioecological impacts of companies. Underreporting combines with abstract indicators and inaccurate measurement devices and, most of all, bounding company accountings to national borders.[72]

While carbon, through another mathematical calculation, represents six elements (nitrous oxide, hydrofluorocarbons, etc.), it does not account for the myriad of industrial wastes, ecological disruptions, and degradations related to techno-industrial development and urbanization. Quality remains a lacking and detrimental factor. Moreover, this does not include the reliability of data, measurement procedures, and models, as with the four collection methods above, there are serious data collection limitations. The domestic material consumption of Co_2 emissions, Marina Requena-i-Mora and Dan Brockington show, is "limited to the amount of material directly used by any national economy," therefore "[i]t does not include the upstream raw materials related to imports and exports originating from outside the national economy."[73] This creates accounting bias that makes high-income countries appear less environmentally destructive. Material Intensity and Environmental Performance Index that monitors material or energy use per unit of GDP, Requena-i-Mora and Brockington also show, "does not consider change in absolute or per capita terms."[74] This "gives the illusion that rich countries can grow indefinitely because natural resources are unlimited and/or substitutable with manufactured capital"[75] and, thus, preserving the myths of the green economy. Any attempts to measure this with any real accuracy—outside (and along with) the narrow ontology and epistemology of modern science—would require time-consuming and costly multi-scientific impact assessments.

Carbon remains a categorical scientific abstraction. This raises issues with how carbon can represent the qualitative dimensions of ecological destruction. Take a mine, for example. Carbon accounting fails to account for toxic dusts entering the air, tailing dams (containing arsenic, thorium or other heavy metals) overflowing or breaking, downstream water contamination, ecosystem die-off, and human rights abuses, enacted to enforce the construction of a mine.[76] Carbon accounting conceals ecological catastrophe. Emphasizing climate and decarbonization remains a sleight of hand, which separates climate from ecological issues. Gelderloos calls this "climate reductionism."[77] This discursive shift, purposely or implicitly, deflects from the reality that climate change is produced by local ecological degradation and around every city, factory, and extraction site. This framework of climate change, meanwhile, sends the paralyzing message that the problem of climate change is enormous, locating action primarily with governments, international committees and companies, which in essence disempowers people from taking immediate ecological action in their localities.[78] This, likewise, creates new industries related to geoengineering.[79] Carbon accounting embodies numerous supply chain monitoring problems, which relate to a large range of governance issues and techniques to avoid accountability. We must, however, remember the ideological systems celebrating capitalism and an intensification of business as usual—accelerating socio-ecological and climate catastrophe—that find carbon as a suitable metric to allow the killing of ecosystems, but in a sanitized, distanced and planetary framing.

Carbon acts as the primary indicator of climate catastrophe, which includes its algebraic and econometric accounting procedures. The reductive approach of carbon, then performs three functions. First, it allows the minimization of ecological and climate catastrophes, providing conservative assessments regarding the impact of techno-capitalist development. Second, it enables half-hearted mitigation schemes, such as conservation "offsets," ecotourism and low-carbon infrastructures, which collaborate with mining companies and further integrate ecosystems into economic and financial circuits. Thirdly, it enables the gymnastics of statistics and modeling, which offers reassurances, but also distracts from questioning the existing intensifying trajectory of production, consumption and profiteering. Carbon accounting, then, emerges as an essential technology in legitimizing the socio-political catastrophe. The reassurances are manipulative, justifying the continued coercion of extractive capitalism and aiding and abetting planetary degradation and destruction.

LET THIS PERCOLATE

The ontology and epistemology guiding techno-capitalist societies raise serious concerns. Modern science, and its materialist ontology, denies the existence of life as inert material to be killed, stored, and transported to serve as material and energy to build and power capitalist societies. Quantification is painfully seductive to government and industry, creating possibilities and justifications, and done so with an aura of authority to convince the public. Yet, as we will see in the following chapters, this scientific approach and governmental imperative comes at a great cost. "[W]hen social phenomena are represented numerically," Trenholme Junghans reminds us:

> context is stripped away, complexity disappears, and nuance and particularity are lost. Not only are the resulting representations of complex social phenomena qualitatively denuded, they also have pernicious staying power, displacing the complex entities and phenomena they are meant to describe, and giving rise to new entities, categories, and standards against which measures and rankings will be made going forward.[80]

The quantitative epistemology associated with modeling, and guiding public policy, has serious consequences in matters of socioecological and climate catastrophe. This scientific approach to ecosystems bears responsibility for the present socioecological and climate catastrophe, as well as the so-called "green," "clean," and "sustainable" development projects they justify. The concerns and contestations outlined here should foreground the rest of the book. The inhabitants of the Isthmus of Tehuantepec, in Oaxaca, were among the first to alert the world of the reality of wind energy development and its socioecological costs, which the so-called "Global North" and urban populations of the world are quick to ignore.

2

Grabbing Istmeño wind: energy colonization and resistance in Oaxaca

Zapotec, Ikoot, and other Istmeño people were among the first peoples to alert the world to the true ecological and social impacts that wind turbines can cause. The Isthmus of Tehuantepec region, or *El Istmo* (the Isthmus), is a five-hour bus ride south of Oaxaca City, Mexico. The Isthmus remains the narrowest point between the Pacific Ocean and the Gulf of Mexico, which is home to a particularly unique geography. Mountains scatter across and mediate the lands between the Gulf and Pacific, meanwhile the region is home to the Lagoon Superior and Inferior that mixes fresh water with the Pacific Ocean (see Figure 2.1). The overall topology, replete with numerous ecosystems, creates an atmospheric pressure and differentials that produce a powerful north wind that blows through the region. This notoriously powerful wind, however, was not recognized as valuable—or profitable—by the Mexican government and transnational until the 2000s. In 2003, a USAID-sponsored report, *Wind Energy Resource Atlas of Oaxaca*, mapped and confirmed this "excellent wind resource."[1] This let the World Bank's financial arm, the International Finance Corporation,[2] to declare that the region has "the best wind resources on earth." This eventually resulted in a "wind rush" in the region, starting in 2006 and continuing into the present.

The Isthmus region is historically a rebellious region, some say the first to begin the Mexican Revolution decades earlier,[3] and in 1981 Juchitán itself the first left wing municipality in Mexico with members from the Isthmus Coalition of Workers, Peasants and Students (COCEI) taking power. This win broke the political

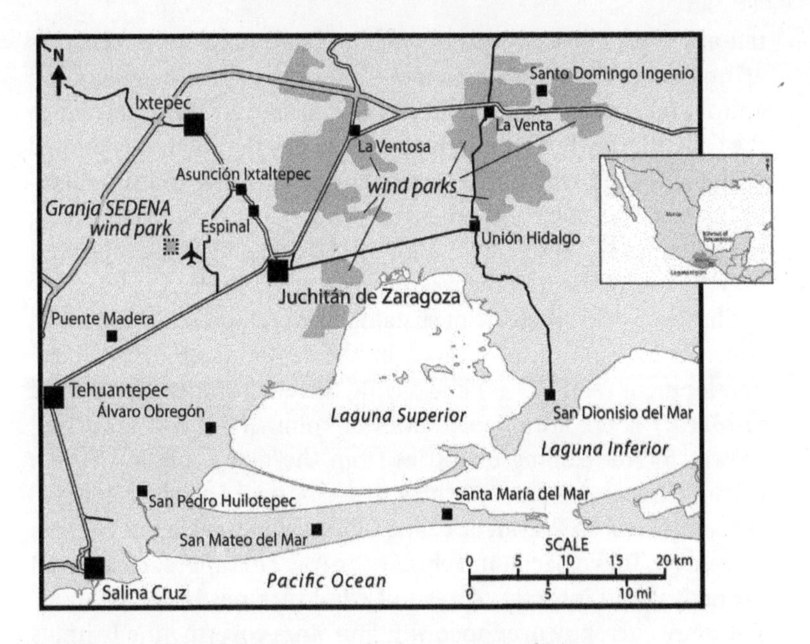

Figure 2.1 Map of the Isthmus of Tehuantepec and Mexico.
Source: Carl Sack, CNS.

dominance of the Institutional Revolutionary Party (PRI) since the end of the Mexican Revolution. This was not without violent struggle and, in 1983, it turned to a military occupation of the city, which ended when COCEI and PRI agreed to co-govern the municipality.[4] The COCEI, and Istmeño struggle, was also about negotiating the influence of the Mexican government and megaproject development. This included the Tehuantepec railway (1907), the Pan American highway (1948), the Benito Juárez Dam (1958), the Trans-Isthmus highway (1960), the electric plant (1960s), the rice-processing plant (1960s), the sugar mill (1960s and 1970s), the Pacific Coast PEMEX refinery (1979), and many other projects.[5] After the 1994 North American Free Trade Agreement (NAFTA), which also triggered the Zapatista insurrection in Chiapas, this legislation permitted greater influence from foreign transitional companies. The Isthmus, however, has remained proactive in defending the region against extractive development projects, which includes successfully resisting eucalyptus plantations, steel mills, industrial shrimp farms, and other schemes arising from Plan Puebla Panamá/The Mesoamerica Integration and Development Project.[6] Regional elites, even those related to the COCEI, slowly began softening their position on transnational influence. This was noticeable with the arrival—and/or toleration—of a Wal-Mart and other transnational superstores entering the region in the early 2000s and, most of all, embracing the wind energy development in the region.

The "win–win" rhetoric of sustainable development had proved effective in gaining a foothold in the region—achieving social development (and riches) and saving the environment with wind turbines. "There are opportunities of commercial and industrial growth for developing countries from the new economy, which will be a green economy … which would be based on clean energy," explains Mexican President Felipe Calderón at the United Nations University Tokyo, Japan in February 2010.[7] This approach allowed the arrival and intensive spread of wind turbines in the Isthmus. There are currently over 2000 wind turbines covering the Isthmus region.[8] This, however, was not without its discontents and eventual ardent resistance to them. It helps to think of the coastal Isthmus divided between the north and the south—or more accurately east and west—but from the perspective of looking at the map you have north and south of Juchitán. The north, primarily Zapotec, allowed the wind rush to take hold between 2005–2011. In 2006–2007, there were contestations over manipulative and disadvantaged land contracting procedures that resulted in a legal win, yet people in the north still proceeded with wind energy development. Development in the north, however, would generate discontent related to failed promises of social development, exploitative (and racist) land deal terms, and ecological impacts. Regarding land contracts, the landowner, "Fernando" collaborating with the wind companies in La Ventosa describes how "we wanted to inform ourselves" before signing any contracts, "so we sought an adviser and it was inconvenient for them, because once we tried to formally associate ourselves with a civil organization and they did not want that. They refused to negotiate with an organized group. Like they say, 'divide and conquer.'"[9] The strategy throughout the region has been and remains a process of "divide and conquer," yet militant resistance against the wind extraction did not begin until the wind turbines threatened the Lagoon and the Barra de Santa Teresa (the Teresa sand bar)—the narrow strip of land separating the Lagoon Inferior and Superior.

Figure 2.2 Northern Isthmus, Asunción Ixtaltepec, Oaxaca, Eólica Sureste I, Fase II wind extraction zone. Source: *Wikicommons.*

WIND ENERGY DEVELOPMENT IN THE ISTHMUS: NORTH AND SOUTH

When discussing the coastal Isthmus, it is useful to think of it as two sections: the north and the south. Sitting at the base of the Atravesada mountain range, the northern part of the region is generally regarded as Zapotec (Binníza), while the southern side is predominately Ikoot (Huave) territory. These territories overlap and are home to five different ethnic groups—Zapotecs, Ikoots (Huave), Zoques, Mixes, and Chontals—and people with mixed ancestry (*mestizo*). Since 2004 wind energy development has penetrated the region in full force, resulting in the construction of over 2000 wind turbines, with 31 wind parks and double this planned for the region.[10] The desire for work, social development, and prosperity is what created a foothold and support for wind projects among the local population in the region. This relates to struggles over the cultural conceptions of land, which is a conflict going back to the Mexican Revolution—if not earlier—and continues with the arrival of wind turbines.

Ejidos, an agrarian concession put into Article 27 of the Mexican constitution after the Revolution, placed abandoned or foreign-owned land into the hands of people under communal tenure regimes that are organized by associations, known as agricultural communities (*comúnidades agraris*). The ejido system also existed alongside communal land (*bienes comunales*), which are recognized Indigenous lands traced back to pre-Spanish occupation. Together, these collective agrarian regimes are called social property. Oaxaca state has 76% of the land classified as social property—"851 ejidos and 719 agrarian communities."[11] Different conceptions of land-use, and ownership regimes remain a central source of disagreement between the Zapotec elite (and caciques) and other collectivized regimes rejecting formalized private property. It is best to situate this modern disagreement with the presidential decrees in 1962 and 1964 that recognized 68,000 hectares of communal land. These decrees, however, are surrounded by institutional ambiguity and contestation by landowners and capitalists in the region. In 1978, Victor Pineda Henestrosa, or Victor Yodo, who was the last "agrarian promoter" (*promotor agrario*) responsible for protecting the communal lands, was abducted and disappeared. This led the communal land "assembly system to collapse"—fueling further legal ambiguity and contestation.[12] The position of agrarian promoter remains vacant to this day, which wind company land contracting and notary practices have only complicated further. Oppositional groups view this land contracting as nothing short of land grabbing, transferring communal land to private property. As the famous quote by the land defender, Wild Tiger, explains, the wind companies

> are taking advantage of mother earth's illness, which is global warming so they can grab all the natural resources from the first nation people (*pueblos orignarios*) from this land. They are grabbing our water, they are grabbing our wind, they are grabbing our lands, they are grabbing our forests, they use the

> protection of natural resources as an excuse; but in Mexico the first nations people are well-known for giving offerings to and respecting the life of mother earth—taking only the natural resources that they need.[13]

Different ontologies and relationships that manifest into different conceptions of land-use regimes underline the wind energy conflict and, as Gerardo Torres Contreras reminds us, the wind turbine "expansion plays a key role for landowners by establishing a valid claim over land."[14] The conspiracy between the "state–capital–property trinity"[15] collaborates to smother alternatives to modernist development and land relations to impose high-quantity of industrial-scale wind turbines for capitalist investment and profiteering.

The towns in the northern coastal Isthmus, La Venta, La Ventosa, and Santo Domingo Ingenio, initially embraced wind energy development. "I analyzed the potential of my land and there was grass on the land and let's say, that grass could get me 2,000 pesos a month," explains a landowner in La Ventosa.[16] "And what they [the wind companies] are going to give me is a lot more." Fernando, another landowner, recounts the benefits of payments being able to send his kids to college, but also notes an element of security with the wind turbines: "It is hard to steal a wind turbine, but it's easy to steal a cow."[17] This same landowner, however, felt "a complete lack of knowledge" concerning the ecological impact of the wind turbines and desired an improved land deal, feeling as though the company was taking advantage of them and that the people in the Isthmus were treated as "second class people." "It is not fair that there are classifications, because the skin color of human beings should not make a difference for us—the mandate is universal and so should the [company] payment[s]." In La Ventosa, a town in the northern coastal Isthmus, many people felt the promises by the wind companies remained unfulfilled, limited, and benefited a minority of the population.[18] The more common phrase is that the wind projects "have absolutely no benefit for this town," which continues with the lack of paved roads, sewage, and improvement in schools and "the town cannot grow and the land is so expensive people cannot buy their land because there is no work or money."[19] This references the boom and bust cycles of temporary jobs working in wind parks, but also the experience of "rural gentrification" that was trigged in the area. The new gentry, in this case the high-income wind company managers, skilled workers, and visiting investors, which as a general influx of foreigners, tastes, and habits—along with the wind turbines themselves enclosing entire towns—would drive up the prices. This benefited local restaurants and landowners, but reinforced existing inequalities and promoted uneven development. Existing landowners, capitalists, priests, and politicians—as they intersect—are positioned to profit from wind energy development, while landless workers and small-scale farmers are largely excluded and are subject to price increases—"the rich get richer, and the poor get poorer," as the saying goes. Wind energy development created a "trickle-down effect" based on inequality and, in line with project construction, was temporary.

Rural gentrification combined with skilled and specialized workers brought "man camp" dynamics to the village. The influx of imported skilled workers, as well as unskilled workers looking for jobs, generated greater demand for drugs, sex work, and crime.[20] Residents, even some collaborating with wind companies, felt the wind companies were invading them. This was reinforced by patriarchal claims that "tall," "white," and "blue-eyed" foreigners were stealing their women and daughters.[21] This invasion also combined with the distance of wind turbines from people's houses. "Yeah, you wake up from the sound of the propeller and a vibration—buzzzzzz—it is bothersome," explains a mother. "When there is music or TV on you cannot hear it, but at night when it is quiet that is when you hear the buzzzzzz and the propeller buzz."[22] This also includes ecological concerns. A wide age demographic was worried about the loss of habitat for iguanas, rabbits, mountain lions, and birds. An interesting phenomenon is the observation of wind turbines leaking oil, which a maintenance worker admits: "Now you can see it. Several of the turbines you can see have oil leaks. If you want to go out in the sun you could see several—30 or 40% are leaking oil."[23] The landowner, Fernando, also observes this:

> The blades are quite stained with grease and you can see that the grease is getting into the ground and that is where I have recommended to my countrymen that they should insist that grease should not be allowed to hit the ground, because it gets into the ground and in a while it will cause problems.[24]

This concern of oil leaking into the ground was mentioned in numerous interviews,[25] confirmed by other researchers[26] and reemerged as a concern in old wind parks in southern France, which we visit in Chapter 5.

The ecological impact of wind turbines is surprisingly extensive. The resources required to construct a single two-megawatt wind turbine are roughly 150 metric tons of steel for reinforced concrete foundations, 250 metric tons for the rotor hubs and nacelles, and 500 metric tons for the tower.[27] This also includes 3.6 tons of copper per megawatt.[28] Furthermore, industrial steel production is impossible without burning coal, as metallurgical coal—or coking coal—is a vital ingredient in the process.[29] While this material cost of wind turbines necessitates intensive mining, chemical processing, manufacturing, and transportation, on sites of development it requires habitat clearance, widening, and fortifying existing roads for heavy machinery or building new roads.[30] This, moreover, includes wind turbine foundations that are filled with steel-reinforced concrete that range, depending on the geography, between 7–14 meters (32–45 ft) deep and about 16–21 meters (52–68 ft) in diameter. Consider this recount from a *comunero*—communal land owner—"Ta Pedro," who after engaging in a collective campaign of sabotage and resistance, was forced to watch the construction of a wind turbine next to his land. A construction process facilitated by force of arms via police, private security, and extra-judicial gunmen.

Driving down the dirt roads into Gas Natural Fenosa's recently completed Bíi Hioxo wind project, it was the first wind project built on the Lagoon in 2014 (e.g., the southern Isthmus). Driving by the energy transformer inside the wind park, we passed the Auxiliary Banking, Industrial and Commercial Police (PABIC)[31] pickup truck, with a police officer armed with an IMI Galil assault rifle in the back. The police stared at us with stern, suspicious, and intimidating faces. When we arrived at Ta Pedro's land, he walked us around. He showed us the oil in his water well, discussed issues with his cattle—he suspected was related to leaking oil—and the different strategies they used to resist the wind park. Eventually, he described how highly armed men began facilitating construction, "they paid some *pistolaros* (gunmen), they only have high caliber weapons—AR15's, big weapons—and we could not do anything about it because we only have stones and what can we do with those?" The Bíi Hioxo Wind Park was built using public and extra-judicial armed forces to impose the project. Pedro points to the wind turbine 30 meters from where we are standing, describing what he witnessed while they built its foundation.

> They brought a lot of machinery, those for digging, they made a ravine and a square that was 20 by 20 meters and it was 12 to 15 meters deep. From where we are standing, if you dig a meter deep you can find water [below the ground] and when they [the company workers] dug those big trenches all the water comes into my land. So that is where the water rises and sometimes the water was green, because they cannot stop the water and my land was flooded and nobody gave me anything for that damage. After they brought some fluids and they poured them into the water and I do not know what happened, but afterwards the water stopped.[32]

Traditionally, the Isthmus has abundant ground water, river, and sea resources, typically finding ground water between 1 to 3 meters below the ground in the region. The rhetoric of "green," "clean," and "renewable" energy often requires excavation, chemical solvents, water table disruption, and, in the southern coastal Isthmus, armed forces. This process of excavation and water table management described by Pedro, we must remember, was done for all 116 wind turbines in the Bíi Hioxo project.

Back in La Ventosa, I walked through the streets, talking haphazardly to anyone who would talk with me. Another unexpected ecological concern emerged: rain loss. At the time I thought it was strange that people between 14 and 70 years old would repeatedly mention rain loss in conversations. I was sympathetic, but unsure how wind turbines would affect the weather. The Zapotec perspective shared with me, while simplistic, was rather convincing.

> The wind energy parks are going to affect everything because it will affect the birds. Where are the birds going to hang out? The tradition here says that when the birds move they ask for the rain, but if those animals die, who is going to ask for the rain? That is how it [the wind turbines] is going to affect [the Isthmus].[33]

Wind turbines, as is well-known, kill large quantities of birds. The blades can hit them, but more often the movement from the turbine makes the birds lose control and send them into the ground. This is in addition to the impacts of high-tension power lines and transformers, which we will discuss in Chapter 5. The Isthmus is a well-known corridor for migratory birds, which is among the reasons the World Bank sponsored a report in the region. The research on wind turbine impact on birds was conducted in the La Venta II Wind Park between 2007–2008, finding that migratory bird mortality rates increased by 20 or more birds per installed wind park megawatt per year.[34] The number of birds killed has undoubtedly risen, which raises the general issue of dwindling bird populations. There was also the issue of concealing bird deaths, often by maintenance workers actively cleaning up bird carcasses within wind extraction zones in the eastern coastal Isthmus. This results in complicating how to accurately chart and calculate bird deaths caused by wind turbines or high-tension power lines.[35] The intensive increase in wind turbines after 2007 is compounded by habitat and tree loss that eliminates the homes of birds, insects, and other species.

The loss of rain and the change in weather observed by people in the region is a known impact of large quantities of wind turbines. Weather and climatic changes are further intensified by large-scale wind energy development. Reviewing studies on how wind turbines impact the climate, S.A. Abbasi and colleagues explain that

> each new study is deepening the realization that a) wind turbines do impact the local weather; b) the impacts are neither negligible

> nor confined to the vicinity of the turbines; and c) the possibility of long-term impacts on climate is high but it cannot be said how severe the impacts can be.[36]

The atmospheric impacts can span between 5 and 21 km downwind from the wind extraction zone. Wind extraction creates a "wind velocity deficit" by interrupting and dissipating wind streams, which is causing an increase in heating and climatic changes that alter and degrade the regional ecology. This makes sense. Large steel, mechanical, and digital apparatuses placed in large quantities in areas to extract wind flow currents will generate atmospheric and eventually climatic impacts—depending on the turbine size, quantity, placement, regional ecology, and other external factors. This issue needs to be taken seriously, and it needs to be recognized that this is happening as an accumulation alongside hydroelectric dams, oil refiners, and hydrocarbon and mineral extraction; chemical production; as well as industrial and urban expansion. Wind extraction accumulates alongside extractive industries. Wind energy impacts remain yet another sign that it is crucial to change the extractive socioecological relationships of humans toward socioecological sustainability and—real and genuine—renewability.

This atmospheric—above-ground—impact coincides with the below-ground impact related to concrete foundations, local topology, and geology. The large concentration of wind turbines and their concrete foundations, described by Ta Pedro, are impacting the water table and how freshwater drains into the Lagoon. Wind turbine interventions are creating extreme drying and flooding according to seasons. In the dry season, wind dispersal into the ground intensifies this process, while roads, power lines and turbine foundations are compacting soil and clogging water veins with reinforced concrete and causing flooding. The popular claims that wind turbines are compatible and can co-exist with agriculture and livestock require prudence and caution. This entails inspecting the quantity, size, and placement of wind turbines with the local geology, water tables, and type of agriculture and farming activities. Between dripping oil into the soil and extreme drying and flooding of croplands, this is a rather significant issue—even if seemingly less attractive, or less of a "monster," than oil refineries and open-pit mines—both of which are necessary for wind and solar panel production in the first place. In the Isthmus, the high concentration of wind turbines is resulting in claims that freshwater drainage is obstructed by concrete foundations.[37] "Every time people have to dig their wells deeper to reach their water," the land defender Jose explains referring to drought in the region. When digging water wells, it makes sense that there is no water because there "is over 100 tons of cement injected into the foundations of each wind turbine. Then multiply that by 2,200. *They are creating a barrier of cement that stops fresh water from reaching the Lagoon system, accelerating its salinity and drying.*"[38] The unique geographical features of the Isthmus, because of mapping and wind valuations, have paved the way toward enormous industrial wind extraction intervention with extremely degrading consequences.

The situation in the northern coastal Isthmus is summarized succinctly, while I sit in the backyard of the TEPEYAC Human Rights Center, in Tehuantepec. We walk to the backyard shaded by tamarind trees as the human rights defender Lobos and his colleague pull up chairs for everyone in a circle. Lobos grabs a stick and begins to draw an outline of the coastal Isthmus on the ground. Lobos explains:

> As the [wind] companies began to arrive, the government was not even acting as the middleman; they came directly to the communities and ... went directly to those people to start making deals. When we started to realize what was going on, we were seeing wind projects being built in this part of the Isthmus [pointing to La Venta and La Ventosa on the map]. *The modernistic inertia was advancing until the notion of wind energy fractured the COCEI itself.* So there was no doubt that wind energy was the least damaging of the ways to produce energy—it's better than a refinery, better than a thermal electric plant, it's better than a hydroelectric plant and the argument that these were lands not being used started to spread. (emphasis added)

In broad explanatory strokes, this was how wind extraction zones gained a foothold in the Isthmus. Then, speaking to the first wave of protests between 2006 and 2010, Lobos continues that there were

> outbreaks of protests in places like Union Hidalgo, in Juchitán, in La Venta, in Santo Dominto [Ingenio]. The protests, well, had a large group of people who did not want to sell their land, but at the same time *the majority of people were protesting because they wanted to be paid more for their land.* (emphasis added)

This interest in *environmental justice*—aiming for capitalist modernity with greater cultural recognition, project participation, and distribution of the "costs" and "benefits" of projects—as opposed to *autonomy*—asserting control, challenging capitalist modernity, and creating alternatives to development (e.g., post-development)—remains an enduring tension between the northern and southern Isthmus, even if the two overlapped. Residents in the north were fighting for greater incorporation, as well as for individual and collective benefits. This includes unions fighting for more wind parks, who also criticize the wind companies for the importation of technical employees and unequal pay between Mexican and Spanish workers.[39] This is why Lobos concludes, discussing the northern Isthmus, that "[t]he Zapotecs allowed them to fill their lands with wind turbines and there was some institutionalized violence, some people were persecuted. The government contacted the leaders and those they were not able to buy-off, they persecuted." Resistance in the north was suppressed with money and political violence, which gave way to wind turbines spreading toward the southern coastal Isthmus.

The southern coastal Isthmus

The preference from companies to approach politicians, elites, and landowners—as the three overlap—left many in the northern Isthmus excluded from any monetary benefits from the wind turbines. Corporate funds to municipalities, local clientelism, and token public relations efforts were the main form of inclusion offered by the companies. With the arrival of wind turbines, the majority of the people felt the intensification of existing inequality, increasing prices in the village (including electricity), and the erosion of self-sufficiency (e.g., agriculture, hunting, collecting wood). This insecurity was generated by new land enclosures patrolled by security guards; rising prices, crime, and electricity; as well as the overall psycho-social impact of being surrounded by wind turbines, some as close as 280 meters from people's homes (even if 500 meters was more common). Wind turbine proximity to houses, along with existing factors, resulted in stress, sleeping disorders and other health ailments.[40] Symptoms associated with *wind turbine syndrome*—noise irritation, exhaustion, insomnia, headaches, dizziness, and so on—were a reoccurring theme for people living on the edges of La Ventosa. Energy extraction did not power local homes, nor reduce the electric bills—this was understood as yet another insult and degrading assault. And coastal Ikoots and Zapotecs were watching.

Instead, this energy was for state foreign direct investment and national and transnational corporate profit. Wal-Mart became a principal investor in wind projects around La Ventosa, eventually negotiating a 60% power share, buying electricity "at a price that is higher than wholesale, but lower than retail."[41] As early as 2009, the Federal Electricity Commission (CFE)[42] had signed agreements to export electricity from the Isthmus to Belize, Guatemala, and California. In addition to Wal-Mart, investors in the northern Isthmus wind projects included: Cemex (the world's largest cement producer), Grupo Bimbo (the world's largest food processing corporation), Grupo Mexico, and Peñoles—two of Latin America's largest mineral extraction and processing companies.[43] This trajectory of wind turbine invasion did not stop. Companies had their appetites set on the entire region, this means heading south (or really east) toward the Lagoon Superior, the Santa Teresa sand bar (Barra), and the Pacific Ocean. These towns and communities had taken note and watched what happened in the northern coastal Isthmus. As Lobos, with his dirt map of the Isthmus on the ground, explains:

> Neither the government, nor the companies or even the Zapotecs imagined that when they entered those lands the

> tables would be turned on them. Because the people who live next to the Lagoon—everyone who lives next to the Lagoon has another notion of the land. There is a 1,000-year-old cultural identity and it has a lot to do with the sea, with the wind—this is a sacred place for the people from San Dionisio [del Mar], for those from San Mateo [del Mar], for those from San Francisco [del Mar] and Álvaro Obregón. They used to come here [pointing to the Santa Teresa sand bar] to do their rituals. So this is a recent phenomenon, it's like an emancipation of the sea peoples (Mareños),* they are sea people because they live next to the sea. Even though the Zapotecs are over here, the Ikoots are on this side and the government never thought—it never crossed their mind that people from Álvaro Obregón would unite with people from San Dioniso and San Mateo—they never thought that. Why? Because historically these people have been passive, they are not a war-like people and the government hit them where it hurts in the Barra de Santa Teresa. They [the government] did not expect this conglomeration of forces. Because the opposition was over here [in the north], people who are politically aware, violent people ... well not violent, but who defended their rights historically—people who are very convincing—and yet they sold their land. When the companies come in here [in the south], there is a different way of thinking.

The attack on the Lagoon and the sea dissolved historical ethnic and existing political tensions to defend the Lagoon and sea from wind turbines. When the wind companies began negotiating with municipal authorities, people were unaware. Then when they found out, the reaction was protest, barricades, and insurrection.

Invading wind turbines means attack

The wind parks and their continued expansion toward the south had become an increasing source of discontent, especially for those who continued to value their subsistence from the land and sea. The specificity of how the wind turbines entered into the differ-

* Another way to say Ikoots.

ent parts of the Isthmus had both difference and commonality, but in the present struggle the north and south coastal Isthmus, as mentioned earlier, represent two different archetypal, yet overlapping, forms of resistance that are revealing of their respective contexts. Opposition in the north, had environmental justice-centered concerns, lacking inclusion and distribution of project benefits, exploitative land deals and labor contracts. In contrast, the southern coastal Isthmus began by fighting for the total rejection of wind energy projects and—in the case of Álvaro Obregón or Gui'Xhi' Ro in Zapotec—Indigenous autonomy. This combative resistance and autonomous struggle was largely fueled by the belief that the wind companies, politicians and the political system itself cannot be trusted, recognizing that party politics only propagates deception organized to grab their land, sea and, together, their ability to subsist and exist in the region. This struggle had an *anti-political*—not to be confused with being apolitical and technocratic—character attractive to anarchists and autonomists.[44]

The south coastal Isthmus, listening and witnessing what happened in the not-too-distant north (only an hour bus ride away), eventually had to confront politicians and some landowners, interested in negotiating the terms of the Santa Teresa sand bar (Barra) project, which is communal land. The Barra is traditionally recognized as common Ikoot territory (meaning it was shared by all neighbors) and officially belonged to the municipality of San Dionisio del Mar. Gui'Xhi' Ro (Álvaro Obregón), a predominately Zapotec town, was the town situated at the land entrance to the Barra. Like in the north, the process of bringing wind turbines into the area began with the help of local elites, politicians and interested *comuneros*. Public consultation was bypassed, with companies opting for selective negotiations with regional administrators, elites and social property members. The general public was left without information, meanwhile politicians began their approach. Referring to Gui'Xhi' Ro, the land defender "Tortuga" remembers when

> The people of the town realized what the hell was going on. At that point they [wind companies] wanted to come in, nobody realized that those guys [wind companies] were bringing their

> stuff [e.g., trucks, heavy equipment, security], they started building a tower on the Barra, and they built a tower, they had started from top to bottom and they finished it and there it was. So then they created a checkpoint, which we had to pass through to get where we always go—along that straight highway and right where it curves they put up a checkpoint. They did not want the fishermen to go there, the state police were set up there, and they did not want the fishermen to go fishing anymore—what do you think about that?! They wanted to order us around. When those guys came back, there was about four of them, they told us, they told people and that is when the conflict started—that was it really. Right away we went to battle and fucked them up.

Tortuga refers to the arrival of the Marña Renovables project. Barricades and violent conflict were also engulfing San Dionisio del Mar, with intense fighting, barricades, shooting, and, even, occasional murders related to this project or gaining political authority.[45] Elections were highly contested and political procedures were blocked by land defenders. The companies entered the area through politicians and the municipalities—and the companies had the money to do so. Politics itself—the science and art of governing people and territories—and its relationship to extractive development, revealed the political system as a structure of socio-ecological conquest—anti-politics emerged as a necessity. Land defenders, then, had to extend their struggle to political autonomy to reject the entire political process based on extractivism.

Social conflict spread across all the southern towns, notably San Dionisio del Mar, Álvaro Obregón, San Mateo del Mar, and Juchitán between the years 2011 and 2015, a struggle that continues in varying intensities today. After violent barricade fighting in 2014, San Dionisio del Mar was occupied by 500 federal police, among them the State Agency of Investigations (AEI), in an attempt to re-establish elections and political control in the town. It eventually took over two years before an election could take place. It took over ten years for the first free, prior, and informed consent (FPIC) consultation to arrive in Juchitán, which—unfortunately—was not a serious process. The FPIC consultation reinforced state-corporate power, meanwhile ignoring local cultural norms, serving as a marketing platform for wind energy companies and, overall, creating the illusion of real information, dialogue, and, by extension, democratic decision-making.[46] The FPIC procedure lacked the information people were eager to learn regarding the specific social, ecological, and health impacts that wind turbines would create. The issue—and extent—of health impacts from wind turbines remains debated and, more so, opens up issues of science and what types of science matter.[47] From this research, I learned that biomedicine and statistical accounting (related to registered biomedical diagnoses) are insufficient to account for the health impacts of wind turbines. Medical pluralism, drawing on Traditional Chinese, Ayurveda and Indigenous medicines,[48] remains fundamentally important to understanding the accumulation of health issues in areas inundated with energy and other extractive infrastructures.[49] Different medical traditions, and onto-epistemologies, can offer greater—and more detailed—insights into how the arrival and operation of wind turbines, or any other extractive project, can cause different disorders and illnesses within people. Medical pluralism is the belief in using the strengths of different sciences together for understanding issues and working to heal and prevent them in the future.

The concealment and lack of information was an issue across the entire Isthmus, yet the key difference in the south is that these were fishing communities dependent on the sea, not only for material, but also for spiritual sustenance—identifying as part of the sea. Speaking in more practical terms, "Grandpa Kitty" implicitly referring to the communitarian police protection of the Barra explains:

> the sea is like a bank for this town and now that we are keeping out the fishermen and wood cutters that caused the conflict. That is when we had to fight and defend the road to the Lagoon, because that is where we live. Our parents brought us up there, we want to bring our kids up there, they lived there, they worked there, and they fished there. That is why we are not in agreement and we will never be in agreement.

Grandpa Kitty, stresses the generational importance of the seas. The Barra remains without wind turbines, even if political turbulence persists, and, meanwhile, the Bíi Hoxio wind park was being developed alongside the Lagoon and on top of sacred sites. Access to historical sites commemorating resistance to the Spanish, routes for religious pilgrimage and ceremony, such as chapels and tomes located at the Guze Venda, Guela Venge, Chigueze, and 12 May in La Chaxada, became sites where access was prohibited by armed guards and, later, occupied by wind turbines. Land control in the service of wind energy extraction retains unforeseen cultural impacts and ontological clashes, which modernist development tends toward subordinating, if not erasing in the long term.

The first wind extraction project proposed for the Lagoon was the poorly named the Marña Renovables project—the "renewing the sea people" project. The Barra comprises primarily sand and vegetation and the Marña Renovables project sought to build 102 wind turbines on it, with another 30 wind turbines in San Mateo del Mar. Because of its geological composition, people speculated the foundation depth for wind turbines on the Barra to be up to 70 meters deep, as opposed to the average 8 to 13 meters deep on the land in the north.[50] Recent testimonies from workers drilling for bedrock on the Santa Teresa sand bar in 2010 attest that bedrock depths range between 17 and 48 meters (56–157 ft.).[51] The first attempt at building a foundation, that Tortuga described above, resulted in the mass killing of fish as far as the eye could see.[52] One fisherman, remembering the drilling on the Santa Teresa sand bar, recounts how the fish "did not die all of a sudden, like a heart attack, they lost their senses and their brains" and "would wiggle and spin in circles in the water without being able to swim straight. Then waves would wash them onto the shore."[53] In addition to the 102 wind turbines, a dock for transporting the wind turbines, a less than 1-kilometer submarine transmission line and a 52-kilometer transmission line to the Ixtepec substation would be built. Residents witnessed how the construction, vibration, and noise from the wind turbines would affect—or really disorient and kill—the aquatic life they depended on for subsistence. Political corruption, unequal land deals, and a loss of access to the sea would combine with fishermen witnessing the mass killing of fish with the pilot wind turbine construction, leading villagers to unite and rise up against the wind companies and political parties. Gui'Xhi' Ro began a process of revolt in 2012, kicking out the company and fending off the police. Skirmishes between state police and the town lasted until "the Battle of the Barra" on 2 February 2013, which inaugurated a process of struggle for land, sea, and political autonomy.

The Battle of the Barra took place on 2 February 2013, when 500 police attempted to invade the city from two different entry points. Tortuga, remembers fighting on the beach near the cemetery, they recount:

> We have to fucking fight—fight time. We cracked some of the state police skulls. They got a good beating, like this—bam! And all the stones thrown and shot, we ran out of stones and on the beach we started using the shells with sharp edges, but we used everything against them. I grabbed the shell, even with the sand in it and I would shoot it at them—boom! I would grab the sharp edge and it would cut me, but I did not care. I was just making mayhem—Boom! Boom! Boom![54]

The Zapotec land defenders would use rocks, slingshots, sticks, fireworks, and *ejido* shotguns to defend themselves. The entire town, even those working for the wind company, rose up to defend the Barra. This was an enormous victory, which eventually turned into a process for political autonomy against the state and companies.

On 8 December 2013, residents would organize at General Charis's old house. Charis was the founder of Álvaro Obregón, naming it after Mexican President Álvaro Obregón who was a companion and friend during the Mexican Revolution. Charis house was a symbolic historical site, which is where the barricade to protect the Barra was initially erected in February 2013 and, subsequently, was transformed into a site for congregation, discussion, and decision-making. That December, people amassed and eventually marched down to the town hall. The group resisting the wind turbines would become known by this time as the communitarians, and on that day peacefully took over the Álvaro Obregón town hall—their strength and combative force had already been demonstrated with the anti-wind extraction uprising. This began a process of attempting to assert autonomy, deciding that they were not going to let "the politicians give orders here anymore—it is time for us to give the orders."[55] People recount over 1,000 people marching with the communitarian police (*Polícia Comunitaria*) leading the way. The communitarian "El Vato" recounts:

> ... we walked over here and we stopped before we got to the town hall [in the town square]. There were police guarding the town hall and we talked to them and we said: "You know what? We are going to be taking on this work guarding this building now. We think that your year of service has now ended. You should just step aside and let us take over." And they said: "That is fine we are leaving." They grabbed their stuff and they left and we stayed here. [We asked each other:] "Now what do we do? What about the [police] patrol vehicles? Let's go talk to the [former] mayor Ricardo and ask him to turn the vehicles over to us officially and voluntarily and also the ambulance and everything else. Okay, let's go." Some of the folks went in like a delegation and when they talked to him, right then he created a document and signed it, turning over the vehicle voluntarily.

While the takeover was peaceful, tensions were high in the village. Although the town could agree on protecting the sea, not everyone could agree on this political process—maintaining sympathies and benefits from the previous municipal regime that was ousted. And, of course, the communitarians remained distrustful of the political motives of politicians and the organization of the political process itself. This led to the formation of the Constitutionalists, which served as a counter-force to the communitarians and, with the support of the Juchitán administration, were negotiating with the wind companies officially and unofficially. The conflict between these two factions—pro and anti—wind turbines and politics, gave way to a low-intensity civil war within the town, one-side backed by politicians, the other by peace groups, anarchists, and autonomous fighters. Numerous battles unfolded, usually around election time, resulting in countless gun wounds, beatings, acts of intimidation, and death.

Meanwhile, the Asamblea Popular del Pueblo Juchiteco (APPJ) was formed on 24 February 2013, which first action began the next day by erecting a barricade on the highway to Playa Vicente to halt the construction of the "Fuerza y Energía Bíi Hioxo Wind Farm." This wind extraction zone was located next to the Lagoon Superior. Before the APPJ, the struggle began as night sabotage against equipment and construction material, but led to the increase in police, private security, surveillance equipment, and gunmen (*pistoleros*) securing the operation on communal land. This struggle transformed into creating a barricade to stop park construction, which lasted until 26 March 2013. On this day 1,200 police came to break the barricade, which resulted in the widespread mobilization of Juchitán, notably the Seventh Section neighborhood. Fighting would last all day and into the night, which included many injuries on both sides, a hostage exchange between police and protesters and a media campaign to slander the land defenders. The struggle against the Bíi Hioxo extraction zone was intense, resulting in large-scale public relations activity, token social development, and repression by police and more so by extra-judicial forces engaged in campaigns of intimidation, assault, and murder. The violent repression, as referenced by Ta Pedro above, was instrumental in enforcing the enclosure of communal land, the degradation of sacred Zapotec sites, clearing habitat, filling the veins of the earth with concrete, and further disturbing marine life with wind turbine construction. The aircraft warning lights that flash and strobe on top of wind towers, as fishers from the Seventh Section neighborhood alert us, further push the fish populations deeper into the Lagoon. Wind turbines further stress the marine life used for subsistence, which results in people having to drive to other towns to fish and this has fermented intercommunal conflicts—especially with people from towns who collaborate with the wind companies. The Communitarian Police patrols I joined in Gui'Xhi' Ro were dedicated to stopping this, in addition to other people poaching turtle eggs from outside the village. The wind turbines, said simply, have had far-reaching socioecological impacts, degrading ecosystems and fermenting violent conflict.

As of 2019, the death count in the Isthmus related to wind energy development was: one "wind farm employee was shot in the head and killed" near La Venta; a member of the APPJ, Hector Regalado Jiménez, was assassinated in 2013; a Constitutionalist police captain was killed during an electoral shootout in Álvaro Obregón in 2016; Alberto Toledo Villalobos was run off a road and beaten to death in 2017; and recently a communitarian police member in Álvaro Obregón, Rolando Crispín López, was assassinated, while an 8-year-old girl was wounded in the process.[56] Accounting for murders in the Isthmus, however, is increasingly difficult, not only due to the way political power and elections relate to extractive development, but also with the rise of narcotic trafficking-related murders in the region. While conducting research on the Isthmus in December 2020, for example, the Jalisco New Generation Cartel (CJNG) sent a series of text messages to nearly everyone on the coastal Isthmus. The message declared a 10 pm curfew and demanded that protesters take down road blockades, stating:

> We are going to deactivate any blockade in the area, and clear them out by any means. To all of those persons involved in blockades, especially in the Zanatepec–Niltepec blockade areas, we are giving notice to withdraw as soon as possible, you are affecting our operations. We have orders from our employer to raze everything on our arrival, we won't be held accountable, we're going to wipe out everything, we don't give a shit if it's children, women, or any asshole walking around acting like a dick, there is a reason why we are the bloodiest group.[57]

This "wiping out" is a threat of massacre, which happened across the Lagoon in San Mateo del Mar. On 21 June 2020, two women and 13 men reportedly associated with disputing elections were brutally murdered in Huazantlán del Río. This massacre is undoubtedly related to elections, maintaining political power, legal and extra-legal commerce, and megaproject development in the region, yet its specific relationship to wind extraction remains only one among many factors, even if a strong factor.[58] Speaking about the distribution of money by companies, social development, and people believing in wind energy, an anarchist land defender summarizes the situation: "Whether it is people directly donating money or not, the fact that people directly support these [green extractive] industries and think they are a 'solution' when in actuality they are making Indigenous people shoot each other."[59] The dream of development and modernist enchantments have a way of dividing and conquering people, betraying the land that gives them life—all industrial centers are experts in this. This betrayal, however, is also enforced. Faux climate change mitigation and so-called decarbonization led by capitalist industry and backed by legal and extra-legal armed force, in reality are renewing capitalist destruction, while generating violent conflict and further stressing, degrading and progressively erasing cultural and biological diversity consistent with the trajectory of modernist development.

LAND IS LIFE

The impacts of wind energy cannot be understated or hidden by claims of "green," "clean," or "renewability." The violence of extractive development and so-called renewable energy needs to be recognized for what it actually does to land, seas and the people who inhabit them—human or nonhuman. Recognizing the problems associated with wind energy development can then allow the proper changes to permit socioecological development that actually mitigates ecological crisis and allows for the self-determination of Zapotec, Ikoot, and other Indigenous groups fighting to maintain their culture, cosmologies, and customs. This, however, means recognizing the harms of colonialism and the imperial Christianizing program that has infected the Isthmus for centuries. Colonialism, moreover, combines with the uncomfortable fact that Zapotec elites want private property regimes and are, or aspire to be, capitalists. Being from an Indigenous nation does not mean a commitment to the culture, medicine, and land. Like everyone, originating at some point from a land-based lifeway and culture, this process can be betrayed. This is to stress the importance of Indigenous resurgence and insurrection, but more so practicing different cosmologies, customs, medicines, and self-defense practices that honor habitats, and love and protect what gives people life. This means reviving practices of self-sufficiency, building autonomy, securing subsistence, habitation, habitat defense—the opposite of dependency on the whims and desires of the global capitalist market—and, together, traditional practices. Contrary to the modernist belief that industrial logistics, supermarkets, and urban environments are the providers of life. Instead, we must remember, it is the land and our ecosystems that play this role of material and emotional sustenance—even if the modernist project has reconfigured this relationship and has positioned industrial systems as the source of all life. Ecological catastrophe, and widespread diversity loss, has shown us that the colonial, modernist, and statist project, despite the technological wonders it has produced, has become a threat to all life and the planet. Capitalism and statism is a virus that is infecting the planet and its inhabitants (in various ways), and the medicine for this is bountiful, even if complex (and discussed in the book's conclusion). Yet, for now, we must stop killing trees, poisoning rivers, and working to create rich socioecological habitats—not destroying and exploiting them and their inhabitants. Real socioecological sustainability, reciprocity and renewability should remain the targets, closing down extractive and exploitative supply webs and lifeways. This is a generational task—the same way statism and capitalism was.

Before turning to the Hambach Forest struggle against coal mining in Germany, it should be recognized that the successes experienced in Oaxaca combined militant popular action with legal action. Despite the hardship, conflictive tensions, violence, and murders in Gui'Xhi' Ro, the communitarians—and all who have struggled—have been successful at protecting the Santa Teresa sand bar from wind turbines until the present. Unfortunately, the Marña Renovables project eventually changed its name to Ecólica del Sur and moved to the northern coastal Isthmus where people were not adequately organized to defend their land (instead embracing the project) and, consequently, filling in the land between the La Ventosa and the Bíi Hioxo wind extraction zones. This intensity and density of wind turbine placement and their concrete foundations, as discussed above, has managed to tamper with the salinity of the Lagoons. Less pronounced militant opposition in the surrounding towns and landowners ready to abandon communal land,[60] allowed the wind extraction zone in the Isthmus to expand. Wind extraction coincides with the exploitation of people, which persists and is expanding in the Isthmus. Wind extraction development in the Isthmus not only expresses severe environmental justice concerns, with unfulfilled social development, and collective and individual payments, but these issues are intensifying with an unapologetic continued invasion of more wind turbines, mines, pipelines, and manufacturing corridors to the region. Social warfare divides and conquer tactics persist and adapt, meanwhile the brute force of local and state police, extra-judicial forces (e.g., unions, narcotics groups, and local gangs) and, Mexican President Andrés Manuel López Obrador's new National Guard continue to sow terror and enforce extraction and new megaproject corridors in the name of security.[61] In the end, land defense works, even if is exhausting, politically complicated and fatally lethal. Yet, the struggle continues in Oaxaca and elsewhere.

3

Fighting the worldeater: coal extraction, resistance, and greenwashing in Germany

The memory is still impressed in my mind, standing on the side of the road holding a bike alongside Andrea Brock in the western German Rhineland. Andrea was from a nearby city, and had been doing long-term research in the area for her Ph.D. and, at this moment, we were studying the repressive strategies and techniques employed to maintain and enforce coal mining in the region. The sky was big in comparison to the relatively flat geography with a distinctly rural terrain. The landscape was riddled with monocrop fields, scattered trees, small rolling hills, villages, and overgrown weeds, combined with what felt like an enormous plain with gray clouds filling the air and mixing with steam and emissions from the hourglass thermal power plants, which looked oddly similar in shape to nuclear power plants. The vitality of the landscape felt low and the atmosphere dreary with small villages, wind turbines, power plants, and monoculture fields that were held together by a road system that was uncomfortable for riding our bikes. On most of the roads, cars frequented only occasionally, but when they arrived the cars were driving fast. Then of course, there is the world's largest opencast lignite coalmine—the Hambach mine—that was one of three in the region (including the Inden and Garzweiler mines) and all of which contributed to interrupting the landscape. Reflecting together on riding our bikes around the Hambach mine, Andrea explains:

> You can read about the mine, you can look at pictures or, even, see footage or aerial footage of the mine, which is really interest-

> ing, impressive, and horrifying. But, it is a very different bodily experience to cycle around the mine, especially given the way that RWE, the mine operator, has shaped not just the mine, but the surrounding area and the places mined before and recultivated. Cycling through the recultivated areas is a weird feeling, but also the abandoned villages. Cycling through the villages still there, struggling to survive, is a very bodily and impacting experience that you do not get from just reading or interviewing, you know? You can really feel it. Obviously, the same with the trees that are cut down. Seeing the ecosystems that are destroyed; seeing that hole, the world's largest human-made hole in front of you ... it's all a bit of an emotional and bodily experience, not just an academic experience, you know?[1]

Particularly interesting, and related to the impact of wind extraction discussed in the previous chapter, is the recultivation area that RWE (Rheinisch-Westfälisches Elektrizitätswerk), the Hambach mine operator, engages in to mitigate, or lessen the damage, of coal mining. Mining companies in Germany are legally required under the German Nature Protection Law (Naturschutzgesetz) and the European Habitats Directive to recultivate the mining area, but also to implement additional compensation measures, or offsets (Ausgleichsmaßnahmen). These measures attempt to lend the company credibility and allow them to position themselves as "good" corporate citizens and to further fabricate the idea of "green" mining.[2] This also includes, in this case, the construction of wind and solar projects to further justify this corporate image and soften the jarring and outrageous extractive reality of coal mining in the area.

The enormity of the mine, and the bucket-wheel Excavator 288 (Bagger 288), is simultaneously frightening and attractive for its sheer abnormality, grotesque imposition and mechanical existence (see Figure 3.1). Regarded as the heaviest land vehicle in the world, at 13,500 tons, the Bagger Excavator takes five years to assemble and has a total cost of USD 100 million.[3] The Bagger 288 was specifically built for the Hambach surface mine, which is calculated to excavate 240,000 tons of coal, or 240,000 cubic meters of overburden daily—the equivalent of a soccer field dug to 30 m (98 ft) deep.[4] According to various compiled sources:

> The coal produced in one day fills 2400 coal wagons. The excavator is up to 220 m (721 ft) long (slightly shorter than Baggers 287 and 293) and approximately 96 m (315 feet) high. The Bagger's operation requires 16.56 megawatts of externally supplied electricity. It can travel 2 to 10 m (6.6 to 32.8 ft) per minute (0.1 to 0.6 km/h). The chassis of the main section is 46 m (151 ft) wide and sits on 3 rows of 4 caterpillar track assemblies, each 3.8 m (12 ft) wide.[5]

The Bagger 288 consumes anything in its path—from fields, forests, and villages—as it migrates across the North Rhine Westphalia (NRW) region. This worldeater *migrates*. RWE propaganda claims in public tours, brochures, and promotional videos that "the mine does not expand, it migrates."[6] This, however, is spun by RWE to integrate this migration into its recultivation schemes, contending that: "First came the diggers ... then nature returns."[7] This statement refers to the Hambach mine biodiversity management plan, which—among other initiatives discussed below—includes a large artificial hill covering 13 km^2, with a height of 280 m, and attempts

Figure 3.1 The Bagger 288 at the Hambach mine, surrounded by wind turbines. Source: herbert2512, pixabay.com.

to re-create habitat for a number of (threatened) species. This artificial hill is called the Sophienhöhe, which also serves to dispose of the initial 2.2 billion m³ overburden, or the dug-up ground left behind the lignite extraction, the mine generated in the first six years of its mining operation.[8] The violent existence and intrusion of the Bagger worldeater has not been without contestation and struggle. This resistance—and "attack"—also corresponds with repressive state and corporate efforts to maintain and enforce coal extraction, continue exorbitant energy consumption and, most of all, obtain profit from this investment into the enormous and costly extractive apparatus that is the Rhineland coal fields.

THE SEEDS OF HAMBACH RESISTANCE

The planning and organization of the Hambach mine began in the 1960s, and it started operation in the 1970s. "[T]he mine was approved really quickly in the rush of the oil crisis of the 1970s," explains Andrea Brock, continuing to remind us that there "was resistance from the very beginning with numerous local groups that weren't necessarily really connected" across the region, spanning the Inden, Hambach, and Garzweiler mines.[9] RWE took over the mine in 1978, meanwhile coal mine opposition in the 1970s and 1980s took the form of civil society and church groups forming and spreading information about the mine, voicing opposition in town hall meetings and civil disobedience actions. Coal mining resistance was largely rooted at the local and regional levels. The German environmental and autonomous (*autonomen*) movement was focused on anti-nuclear struggles, engaging in mass protest, physical occupation, vandalism and sabotage of nuclear reactor construction sites, among other actions.[10] There was little oppositional success to coal mining in the first decades because, according to Andrea, "coal was unquestioned, and ecological concerns were not in the media."[11] Coal mining as an ecological issue began to take hold in the late 1990s alongside the rise of climate change concerns, which became stronger in the 2000s.

The climate justice movement further pressed the problem of coal and other forms of hydrocarbon extraction. While there had been steady regional opposition to the Hambach, Inden, and Garzweiler coal mines, Friends of the Earth (BUND) Germany also began to support the struggle in the mid-1990s and 2000s. The environmental direct action movement did not appear until 2010. Between the late 1990s and 2000s, "summit hopping" associated with the anti-globalization movement was a popular practice among social justice advocates, anti-capitalists and anarchist rebels as they would "hop" from international summit-to-summit to protest. The World Trade Organization (WTO), International Monetary Fund (IMF), International Group of Seven (G7), and political party conventions among others were among the targets.[12] When cities announced hosting international summits, local groups would form in those cities and organize counter-demonstrations and events against those public global economic and governmental conspiracies. Counter-demonstrations and civil disobedient actions would often turn into riots and uprisings in an effort to disrupt and shut down the meetings. Anarchist "black bloc" tactics, originating from autonomous groups in West Germany in the early 1980s, that entailed wearing masks and employing mass vandalism, blockades, and property destruction would give the anti-globalization movement notoriety, media attention, and, of course, ridicule from mainstream channels.[13] Critical discussion regarding "summit hopping" was well established by 2009,[14] reflecting on the energy-intensive efforts to organize counter-demonstrations, the abstract nature of protesting large financial institutions, the predictability (and increasing repression and militarization by state authorities) of the protests, and the exhaustive routine of events and legal fees for arrested protesters was being felt. Meanwhile, lasting change from these efforts was also feeling distant. The counter-demonstrations were a bi-annual or one-time "weekend warrior" style event instead of cultivating permanent conflict or nurturing sustained struggles in particular localities or directly with those most affected by the policies of financial institutions—people were rooting their struggle in others, elsewhere, in abstraction and not within themselves to cultivate common interest and complicity in the struggle.

This "summit hopping" dynamic was simultaneously emerging within the climate justice movement. While discord fermented against moderate factions within climate negotiations in 2002, it was not until the 2007 United Nations Conference of the Parties (COP) 13 summit in Bali that led to the emergence of the Climate Justice Now! network.[15] Anti-capitalist summit hopping quickly transferred into environmental and climate politics, which manifested in counter-demonstrations at the COP15 in Copenhagen, Denmark, to place more pressure on global leaders to take climate and ecological crisis seriously. COP15 counter-demonstrations resulted in large-scale demonstrations, media awareness, and struggle, alongside repression with pre-emptive surveillance and detainment, mass arrests and putting people in cages to manage the large influx of arrested protesters.[16] With all this repression, results or "wins" were failing to materialize. It was time to for these movements to change strategy. "After the COP15 climate summit in Copenhagen in 2009," explains Tadzio Müller,[17] "demonstrated to the climate change movement and its more radical climate justice wing that little should be expected from 'the powers that be' in the fight against fossil fuels, they began to focus on local and national energy struggles."[18] Critical self-reflection by environmentalists to create effective strategy, alongside existing ecological anarchist militancy, would plant the seeds for the Hambach Forest struggle to bloom.

After COP15, environmental and climate activists began focusing their efforts on a specific location. Since 2010, environmental and climate activists have initiated a campaign against RWE's three coal mines in the Rhineland. This popular effort, aimed at reaching mainstream audiences, would pick up momentum by organizing climate camps and committing mass civil disobedience actions against the Hambach mine. Large-scale camps and civil disobedience actions would continue, eventually formalizing into the organizing body known as Ende Gelände in 2015 that stands for "here and no further" or "end terrain." While these organizing efforts were essential to organize people and bring attention to the Hambach mine, environmental activists, concerned citizens, and eco-anarchists felt dissatisfaction at what could be called the reproduction of the "weekend warrior" dynamic with bi-annual camps and actions against the Hambach coal mine. This dynamic embodies the modality of political "activism" decried by anarchists and autonomous fighters.

The 1999 text titled "Give Up Activism" by Andrew X annunciates this concern forcefully, reflecting on the Take Back the Streets demonstrations in London.[19] This critique of activism (and activist identity) applies to the environmental movement, which tokenizes and compartmentalizes struggles to particular moments and events, like a hobby and not a lived trajectory consistent with one's life. This, moreover, is compounded by the etiquette of "activist culture" that creates identity, and informal hierarchies and transforms struggle into a job "that fits in with our psychology and our upbringing" and conditioning in market societies.[20] "The activist feels obliged to keep plugging away at the same old routine unthinkingly," contends Andrew X, "unable to stop or consider, the main thing being that the activist is kept busy and assuages her guilt by banging her head against a brick wall if necessary."[21] This reference to routine, exhaustion and "being kept busy" for moral reasons speaks to "activist burnout" and ineffectual feelings related to political organizing. "Activist" struggle, as it is commonly accepted, emerges as politically acceptable, being another identity for self-branding in capitalist societies where one feels a sense of importance, and, most of all, fails to challenge capitalist and extractive power structures. Some people within the environmental movement wanted to break with this "activism" relationship, instead to begin a process of living in permanent ecological conflict—taking pleasure in living in place and committing to permanent conflict against the Hambach coal mine. This means living to obstruct, stop and, if possible, destroy the mine. In 2012, this process was inaugurated with the so-called "occupation" or, more accurately, re-inhabiting of the Hambach Forest with treehouses, shacks, and dry toilets. The process of permanent ecological conflict implied building a relationship with the Hambach Forest, living with it and in permanent defense of it and, consequently, living in offense against the migrating coal mine.

THE FOREST FIGHTS, THE FOREST LIVES

Living with, and even becoming the forest, changes the relationship and dynamic from which one struggles and fights. The relatively token relationship of weekend protests, camps, and resulting feel-good sentiment—not to forget the rise in protest "selfies" with the later rise in "climate youth in 2017"—is immediately tested and put into question by committing to leaving the city, or modernist housing, to live with and co-habit with a forest under immediate attack by RWE with the intention of extermination. The "Forest not Coal" festival, attended by approximately 200 people, became the platform to go from protest events to resistance in the form of permanent ecological conflict.[22] "During a cultural festival in the woods, under the slogan 'Forest not Coal,' activists pulled up platforms in the trees," explains an anonymous land defender.[23] The date 4 April 2012, marked the beginning of living in permanent ecological conflict with the mine. And while the people actively living in the forest initially was in the range of a dozen, there was some active support from the environmental movement considered broadly and locals in the region. Re-inhabiting the forest was an attempt to surpass the "small path of legal protest," recognizing the failure of previous protests but also the absurdity of the law. Land defenders recognized that RWE is "legally allowed" to destroy "the regional basis of living, as well as endangering people's health in the area, not speaking of the world climate,"[24] revealing the purpose and function of the state—to transform living ecosystems into money, urban infrastructure, and electricity. The law emerges as irresponsible and illegitimate.

The re-inhabitation of the forest became a site for organizing events, notably skill shares—how to climb and live in trees, make barricades and so on—more climate camps, and the "UnEvictable-Festival" to gain support under threat of eviction. Eventually in November 2012, 500 police, riot police, and police "climbing squats" raided and occupied the forest to evict people chained or "locked on" to the trees, cubes of concrete, and within underground tunnels. It took four days for the police to evict people chained in underground tunnels, which locals claim is the most expensive eviction in German police history.[25] Not long after the eviction, a local supporter of the re-inhabitation of the forest—and with an oppositional history to RWE—was arrested in the land adjacent to the forest during the eviction. This land became known as "the Meadow." After the release from jail, this resident bought the land, acquiring a permanent legal space to organize against the mine and, consequently, support the inhabitation of the forest. The Meadow would be a crucial outlet for organizing, welcoming visitors, and organizing events, but it was also under constant bureaucratic attack and surveillance from the government and RWE. The Hambach Forest was again re-inhabited on 19 March 2013—a couple of months later—but was then evicted two days later. Eviction resistance, and the civil disobedience approach, entangled activists in detainment, numerous court dates, and legal fees. In short, "lawfare" to slow down, wear down and demoralize land defenders. This, however, did not work.

In July and August, a Climate and Reclaim the Fields Camp was organized in the nearby town of Mannheim. This camp hosted skill-sharing workshops, people through squatting and organized a group direct action to blockade the coal railway. Meanwhile, the Hambach Forest was re-inhabited for the third time. This initiated a protracted struggle, a cat and mouse of destroying toilets, kitchens and tree houses (see Figure 3.2), while land defenders continued to rebuild them and make barricades. According to land defenders, there were twelve police operations in 2013 intent on clearing the forest.[26] This commitment to living with and protecting the forest resulted in tactical innovation, which began with creating objects that looked like mines or improvised explosive devices (IEDs)—with metallic objects with different colored wires coming out of the ground—and posting signs that would say "minefield." The idea was to further obstruct and slow down police invasion, forcing them to call the bomb squad unit to either defuse the object or confirm the contents of the device were in fact fake. Police, of course, would further harass, arrest, fingerprint and occasionally beat land defenders. This would extend to the deployment of police checkpoints, helicopters and drones to monitor the recalcitrant forest. Police and RWE security invasion, violence, and surveillance would lead to escalating conflict. This escalation would be particularly noticeable during "cutting season." This is the time when the forest is cleared between October and February to allow the migrating mine to expand and leave a reconfigured—or reterritorialized—landscape in its wake. In 2013, tactics began to change to stop the mine and counter-police and judicial repression. Civil disobedience tactics, such as barricades, locking-on to extractive equipment, and forest habitation would combine with proactive self-defense against the mine by sabotaging infrastructure, and ambushing RWE security and police from the forest. People broke and stole measuring posts, sabotaging the coal railway and setting fire to equipment. By 1 December 2013, the coal railway moved to another railway track because of "interruptions" and the forest remained inhabited by people living in six trees, a two-story tree house and three regular ones.[27]

Figure 3.2 Some tree houses in the Hambach Forest.
Source: *Leonhard Lenz/Wikicommons.*

RWE'S INFLUENCE, CONTROL, AND PROPAGANDA

Extractive regimes are frequently rooted in settler colonial and authoritative regimes. The organization of socioecological sacrifice—or mining—depends on claims and embodied practices that assume humans are superior to other animals and plants, which can therefore be murdered. This is known as "human supremacy" or anthropocentrism and remains the foundation to pave the way toward imperialistic mentalities and relationships. Asphalt, the toxic hydrocarbon-based substance, paves over ecosystems, enclosing and poisoning them. It is implicitly, or as a matter of fact, deemed more important and/or superior to what existed in its place before, and frequently organized in the service of housing development, market mobility and convenience. The logic of human supremacy over ecosystems and animals, then, spreads between humans with ethnocentric or racist logics that justify and claim the "right" to the invasion and colonization of different people and landscapes. People are then "paved over" or subordinated to plantations, concentration camps/reservations, and other infrastructures and technologies. Patrick Wolfe sees colonial genocidal processes in three non-deterministic phases: (1) the initial confrontation (or invasion), which often includes local elites allowing colonial powers footholds into particular geographies and cultures; (2) the "carceration period" that entails the refinement of population control through displacement, resettlement and infrastructural practices; and (3) assimilation period that aims to integrate Indigenous populations into the colonial system.[28]

Justifications to enclose and subordinate people rest on the real or imagined "difference" that are positioned to make superiority claims, which frequently circulate around cultural, physical and ethnic characteristics. The claim of "better" physical characteristics extends to cultures and ways of life, historically rooted in denigrating land-based cultures, cosmologies and subsistence patterns as "primitive," "backward," and "inferior" as opposed to the "complexity" of industrialization, modernism, and capitalist development. This colonial discourse ignores the complexity of socio-cultural practices embedded in ecosystems, meanwhile promoting racism to justify political rule, subjugation, and ecocidal extraction to power industrial development. Something anthropologists and historians have been historically important in justifying.[29] While in Germany, this relation relates to inter-European colonization that stretches back to the Roman Empire.[30] Coal extraction, however, was elevated to "strategic military status" under the Nazi regime in 1935. The Nazi regime is well-known for bringing colonialism back to Germany to enact a racial genocidal campaign,[31] coal mining was instrumental to that campaign and was elevated to a special status under the Law for Promoting the

Energy Industry.[32] This Nazi law was never eliminated post-war, meanwhile the 1980 Federal Mining Act stipulates the "compulsory relinquishment of private property to mining companies ... by eminent domain whenever public welfare is served, particularly for providing the market with raw materials, securing employment in the mining industry, stabilizing regional economies, or promoting sensible and orderly mining procedures."[33] Mining, as is common in many countries, is then regarded as a "national security interest" to maintain its productive, consumptive and militaristic capabilities. Said differently, coal mining and fascism have an interlinked dependency, the laws of which continue into the present.

The general statewide support for mining combines with RWE's entrenchment into every aspect of social and political life in the Rhineland. RWE is the largest German electricity provider and utility company. RWE historically held monopoly control over the (West) German electricity grid and has consistently positioned the company's interests as synonymous with the national interests of Germany.[34] For example, mapping the relationship between German politicians and the coal industry, Gerald Neubauer reveals how 17 high-level politicians spanning the entire political spectrum retained close ties to RWE as board members, chair and general payroll.[35] This extends to environmental ministers, meanwhile two politicians resigned in 2004 when it was proved they received payments from RWE of €60,000 to 81,000 a year.[36] RWE remains a central pillar within the German political economy, meanwhile working systematically to gain influence and political control in the Rhineland region, which entails exerting significant influence within every facet of society. Municipalities in the Rhineland have significant amounts of power, which has made local politicians win over the "hearts" and "minds" of the general public instrumental to RWE's operations.

Andrea Brock and I have shown how this infiltration of political, cultural, and social bonds is rather intimate, which entails local authorities not only approving RWEs operations, but also acting as RWE shareholders, clients, constituencies, employees, and tax collectors.[37] The conflict of political interests are intense, and become self-reinforcing as RWE has immense political power by funding and controlling infrastructural amenities in municipalities. While counterintuitive, this is actually legal under German law. The law states that politicians can serve on company boards and advisory committees, ignoring potential vested political and financial interests. Politicians, who decide to take a stand against RWE, while unlikely, fear it will have political and material consequences on their careers and families. RWE convenes so-called "regional advisory councils," which in 2004 revealed that the company had been paying over 100 local politicians some €600,000 a year.[38] "'They are everywhere,' explains a concerned citizen referring to RWE, continuing that RWE did well 'to get people into all [the political] positions that matter ...' Wherever decisions are taken, you find people who work for RWE or have worked for RWE."[39] People contend that RWE has infiltrated all of the regional decision-making bodies. Andrea recently reminds us that a quarter of RWE's "shares are owned by towns and cities in the Rhineland, making them dependent on RWE's financial success."[40] This, we could say, is legalized corruption that extends to RWE's corporate social responsibility (CSR) approach, which more or less attempts to sponsor schools, cultural events, sports teams, and literally everything that could create a positive public image, while organizing local dependence on the company for income, subsistence, and entertainment. This was visible within towns, when social events and festivals advertised at bus stops promoted the logos of RWE or one of their subsidiaries. Members of civil society groups describe RWE as an imperialistic "octopus,"[41] seeking to touch and control everyone with its money, infrastructure and jobs.

RWEs influence extends to so-called "private–public partnerships" with the police. It has been proved that high-ranking police were working on advisory boards for RWE, which corresponds to sharing vehicles, personnel, funds, and other resources with the company. The politician responsible for controlling the August 2015 Ende Gelände mass direct action against the Hambach mine, Andrea Brock shows, was a member of RWE's advisory board until just weeks before the protest action.[42] Civil society groups and land defenders in the Hambach Forest also document police and RWE personal collaboration. NRW police force is controlled by the parliamentary Interior Committee and, as Andrea points out,

Gregor Golland who was an MP doubling as an RWE employee with an annual corporate salary of up to €120,000—only slightly less than his government salary of €128.712—and was responsible for internal security and police deployment. This, of course, includes major police operations in the Hambach Forest.[43] The police and RWE security frequently work side-by-side to repress land defenders and environmental activists. This entails beatings, arrests, abductions off the street, attempted vehicular manslaughter by RWE security, the threat of rape, and spraying water on activists locked on to equipment in freezing temperatures. In 2016, freezing land defenders into submission gained notoriety in Native North America, when people resisting the Dakota Access Pipeline were sprayed by police water cannon trucks in freezing temperatures causing mass injury, hypothermia, and near death.[44] Andrea and I have documented that similar political-military, or counterinsurgency, tactics are employed in Germany as in other contexts in Latin America and elsewhere, but with a lower intensity of violence suitable for the German political context. Considering RWE is either directly or indirectly sponsoring local politicians (or their initiatives), in contrast to the mainstream media, it appears there is more credibility to the concerns of forest defenders than someone might want to initially believe.

Greening extraction

RWE has a restaurant and resort-style lookout, called: *Terra Nova*, with sunbathing chairs lined up around the Hambach mine, so guests can catch a tan while watching the gargantuan worldeating mega-excavators create the largest hole in Europe. Similar to the "win–win" rhetoric of conservation and sustainable development projects—like wind energy development discussed earlier—RWE attempts to position the mine as an important investment for the region, providing jobs, sponsoring local events, and donating to municipalities in the region. The social development, or corporate social responsibility plan, combines with ecological initiatives to assure residents and distant onlookers that coal mining—while degrading—is done in the most ecologically responsible and safe way. This is why RWE understands that "recultivation must generally produce high-quality arable areas" and invests in "renewable and storage technologies" that "will be climate neutral by 2040."[45] RWE has invested more than gross €50 billion gross in this decade.[46] As discussed in Chapter 1, the logic of carbon accounting and extractive development assist in positioning industrial manufacturers and extractive entrepreneurial giants as "green" and environmentally friendly. Manufacturing an environmentally friendly profile confuses onlookers, engineers comfort with ecologically destructive mining (and processing) activates, and, more immediately, attempts to undermine the claims of forest defenders and eco-anarchists.

Andrea Brock has explored RWE's so-called "green" environmental remediation and recultivation schemes in her Ph.D.,[47] and elsewhere.[48] Since 2009, RWE has actively been trying to position itself as environmentally responsible with the "voRWEg gehen" campaign, which is a play on words that translates into "leading the way," even when only 2.7% of their energy was based on low-carbon infrastructures,[49] mistakenly called "renewables."[50] RWE has also collaborated with the International Union for the Conservation of Nature (IUCN), the Cologne Bureau for Faunistics, and internationally in the "Bettercoal Initiative." Andrea has criticized how RWE has enrolled two prominent German academic environmentalists from Germanwatch and the Wuppertal Institute in its Corporate Responsibility Stakeholder Council.[51] RWE has gone above and beyond to solicit professionals, accommodate existing German legislation, and sell coal mining—and industrial extractive activities—as "green" and environmentally friendly. This approach also continues with coal supply-web monitoring, which RWE is criticized for "outsourcing" their responsibilities to other companies to avert attention from their destructive operations. In the run-up to the 2013 World Economic Forum in Davos, RWE announced its cooperation with IUCN "for the protection of biological diversity"[52]—meanwhile powering titanic earth-eating machines forming the largest hole and coal zone in Europe.

As mentioned above, RWE has the Sophienhöhe, which also serves as a way to dispose of the initial 2.2 billion m^3 overburden generated from mining. The Sophienhöhe serves as a type of environmental "offset," which claims to mitigate damages from the coal mine in one site by restoring ecosystems in another site by planting trees and supporting biodiversity.[53] The Sophienhöhe contains 150 km of hiking and cycling trails curated by information points with QR codes and offering visitors fabricated lookout points and locations. This includes a Celtic tree circle, for example, and other information about trees, animals, and plants that are viewable from your "smart" phone.[54] Unsurprisingly, RWE ignores the extractive socioecological costs associated with digital and computational devices are now encouraged with QR codes and related infrastructure that are littered across the Sophienhöhe. More still, German law now permits private companies to sell eco-points for recultivating or "improving" landscapes. Eco-points, Andrea shows us, are related to eco-accounts that mimic the concept of biodiversity banks and convert ecological restoration into a banking system. While this might sound positive, it ignores how companies are enclosing and privatizing land, reducing the quality of ecosystems—in this case, a 12,000-year-old forest with young trees on a mound of coal backfill—and justifying not only the mining operation itself, but also the financialization and trading of ecological restoration. Andrea explains:

> RWE's compensation work—the restoration work to offset the impacts of its coal mines and its generation of eco-points—relies on the reduction, abstraction and quantification of forests, grasslands and fruit orchards into "net" numbers and categories of hectares, trees and habitat. These processes of reduction and quantification are the foundation for accumulation by restoration, which depends on the ability to make nature legible in the eyes of economic and numerical valuation, "alongside the institutionalization of the technical language of 'neutrality' or 'net gain' of land, biodiversity and other characteristics and functions of nature."[55]

The ecological devastation of coal mining, then, through digital and financial constructs such as eco-points and banking, allows the accumulation of capital, Andrea and Huff call this "accumulation by restoration."[56] The buying and trading of eco-points facilitates and creates a circular logic that justifies mines, urbanization, and allows bank accounts to grow, meanwhile reductive metrics, discussed in Chapter 1, enable the quality of ecosystems to decline and further separate people from nature—allowing access primarily through digital devices and tours. Mining facilitates an extractive relationship, which becomes reconfigured with digital technologies to expand land control and accumulation by every means possible. This extractive accumulation program, moreover, conditions how people are supposed to live with and interact with forests and wider ecosystems.

The Sophienhöhe combines with other biodiversity mitigation efforts. This includes a network of "bat highways" that are double-tree lines designed to attract favorable insects to bats, which intends to help facilitate the migration of the threatened Bechstein bat. This is done by attempting to create corridors between fragmented bat habitats within the remaining forests surrounding the mine. Within this bat highway is a €4 million "green bridge" over the nearby A61 highway to serve as "crossing aid for the bats from Hambach For[e]st."[57] The Sophienhöhe, bat highways, and other "biodiversity" mitigation measures are complemented by the planting of trees and the construction of solar panels along neighboring highways—demonstrating RWE's commitment to environmental concerns on the highway. This also includes the placement of wind turbines around the mine, which Andrea contends serves "as compensation measures for the mine while concealing the equally, if not more ecologically disastrous mining operations and displacement required elsewhere for the resources necessary for these windmills."[58] This is ironic considering how RWE was documented funding initiatives designed to resist and discredit wind energy development in the 1990s,[59] which has subsequently led to investing in wind energy through their subsidiary company Innogy.[60] This tactical shift illuminates a trend that emerged in the 2000s when large companies began diversifying corporate portfolios and products by embracing and including other energy extractive sectors—wind, solar, biomass, and hydroelectric—along with conventional hydrocarbon extraction. This shift toward corporate integration places greater importance on critically assessing and dissecting claims of "green energy transition," which will be discussed in greater depth in Chapter 5. There remains, however, a general trajectory toward total extractivism, in which the imperative—or mission—of states, economies and companies to maximize economic growth and profit leads toward a competition to attempt to extract everything—ecosystems, animals and humans. While a daunting reality, there are many committed and ardent fighters risking everything they have to oppose this worldeating trajectory in the German Rhineland and beyond, which deserves greater recognition and acknowledgment.

PERMANENT ECOLOGICAL CONFLICT: "THERE ARE NO JOBS ON A DEAD PLANET!"

The Hambach mine has become a visible national issue, gaining a steady momentum of popular attention. Yet, the continuous struggle—and militant attack—remains under-acknowledged. When the Hambach Forest features in academic literature, few if any recognize the level of combative action that took place between 2012 and 2018.[61] The liberal values and vision of academia, then combine with the popular discourse and preoccupation with phasing-out coal. Discussions around energy transition, degrowth, and mainstream environmental movements—related to week or weekend warrior dynamics—have tended to take all of the attention and monopolized the discourse around the Hambach mine, at least in mainstream channels, away from the "teeth" or illegalism in the struggle related to sabotage, vandalism and arson, that has powered this struggle. This tended to perpetuate divisions between protest tactics, and perpetuated the myth of civilized resistance.

Consider Greta Thunberg's visit to the Hambach Forest during the COP23 in Bonn. While Greta was welcomed into the Hambach by most forest defenders—presumably for popular media attention—the Hambach Forest struggle lends credibility, even legitimacy, to Greta's promotion of "climate emergency" at this time. And, as discussed in Chapter 1, the idea of climate emergency relies on the logic of "climate reductionism," reducing complex socioecological issues to carbon accounting and international politicking. This serves green capitalist interests seeking to tap into public funds to advance the green economy and its decarbonization, conservation, and digitalization schemes. The

Hambach struggle against domination, the green economy and its militancy were systematically ignored—and possibly intentionally as a media strategy. It is nice to think, however, of Greta's visit to the Hambach Forest as public advocacy for sabotage and arson in defense of forests, wider ecosystems, and climate. This type of public recognition and support would be a "game changer" if this youth celebrity, taking center stage in the international hallways of governance, would actually condemn green capitalism and promote an ecosystem of tactics to defeat coal (and other forms of mining).

The degrowth movement remains another example. While the degrowth might be the most honest position within academia, as mentioned in the introduction, it has radically lacked a connection to struggles on the ground. In 2020, the first degrowth book, which is coincidently German-focused, attempting to connect degrowth with political struggle is titled: *Degrowth in Movement(s)*.[62] The book discussed community gardens, artivism, Buen Vivir, basic income, care work, and climate justice among others, but completely ignored the self-defense, sabotage, and material attacks against RWE and government forces. Is not destroying extractive infrastructure the purest and most direct form of enacting degrowth? The book did offer a chapter about Ende Gelände that, while insightful, reproduced this general academic tendency to ignore the attacks against RWE and promote civil disobedience. This focus on one tactic—civil disobedience lock-ons and mass occupations—ignores the diversity of tactics that actually win struggles against extractive projects. The mythological history surrounding civil disobedience and pacifism, as Ward Churchill,[63] Peter Gelderloos,[64] Vicky Osterweil,[65] and Aric McBay[66] remind us, ignores the combative—even terroristic—movements that operated alongside nonviolent struggles (giving them bargaining power with authorities because they were moderate enough to engage with to shape political stability). This extends to governments promoting this convenient framing of history and, alongside the US military promoting restricted forms of nonviolence to keep resistance movements' docile, limited and structural change always outside arms reach. The Hambach was a nonviolent and self-defense movement, employing a wide diversity of tactics but among them property destruction, sabotage and physical attacks against RWE security and police when they encroach on the forest or attack the inhabitants. This, however, was a topic of conversation—mostly written on the walls of dry toilets and portable toilets—when the Ende Gelände movement formed, organized actions, and arrived at the Hambach mine in 2015.

By the time Ende Gelände officially arrived in the area, four people had been arrested in April 2014 and one person remained in jail, "Jus." Jus was arrested for defending themself from the police when they were climbing up into their tree house to attack and pull them out and arrest them for being in the forest. During the first Ende Gelände climate camp, they were still in jail. Forest defenders expected solidarity from Ende Gelände: acknowledgment of their situation and political concerns, funds for imprisoned friends or lawyer fees. At the time there was a tendency for the organizers of the camp and its attendees to separate themselves from the forest defenders (and their militancy), and to focus instead on their own actions—the mass organizing and running into the mine to temporarily occupy it and lock-on to the machinery to create disruptions and delays. This disposition generated friction. Yet, importantly, it risks falling into the good/bad protester dichotomy employed by governments and the media,[67] which generated ill will from forest defenders, intense debate and a bitter co-existence between ecological and climate activists, at least in 2015 and after the following year.

Governments and companies, as everyone knows, are always trying to divide resistance movements and protesters. Ironically, this has been done through people—often anarchists and autonomous actors—actually attacking and vandalizing corporate and company property. The examples are endless, from the "Battle of Seattle" (1999) to Occupy (2010–2013),[68] when less determined, or militant, factions feel uncomfortable and mistakenly distance themselves from direct action to fit with the dominant message, concerns, and morality promoted by the media and the extractive profiteers being protested—from Starbucks to RWE. Forgetting the importance of economic damage, delays, and disruptions and condemning militant action creates divisions within movements. Dividing resistance movements through propaganda, capitalist morality, and myth is one way companies try to recuperate damages and losses incurred by resistance movements. Ende Gelände, however, became more conscientious about the problematics of this separation and the harm it was causing within the larger anti-extractivist and climate justice movement. Movements tend to forget that the purpose is to destroy these socioecological destructive operations and resort to in-fighting. This issue, and the fruits of these earlier debates, appears to be blooming currently in January 2023. During this writing, diverse political tendencies, ranging from Ende Geländers, forest defenders, and eco-anarchists—are combining mass protests, lock-ons, vandalism, sabotage, and arson in the defense of Lützerath village,[69] which is the re-inhabited town fighting the expansion of the Garzweiler mine. "Unlike earlier anti-coal protests in Germany," explains Andrea, "there was no condemnation, no appeals for [pacifist] 'nonviolence' or 'peaceful protest.' People have embraced a diversity of tactics, not letting the state and RWE divide and rule."[70]

These tactical issues also connect to questions of political critique and analysis. Hambach Forest defenders and Ende Gelände had always related to industrial and capitalist energy production rather differently until more recently. Ende Gelände, and related movements, have clung to the fossil fuel versus renewable energy dichotomy, which asserts that so-called "renewable energy" is in fact "good" and ecologically sustainable (an idea that invizibilizes their extractive and manufacturing supply webs) and fossil fuels are "bad" and dirty, which is more visibly apparent. Hambach land defenders rejected this dichotomy and marketing ploy.

> RWE-Innogy, the subsidiary complicit in the farce of greenwashing industrialism, which boasts of "renewable" or "sustainable" energy operations. With this approach, RWE feeds the insatiable hunger of the industrial leviathan and at the same time satisfies the manufactured needs of those "benevolent" and "conscious" consumers, so that they too may spend all their lives in the glow of artificial light, with dead eyes staring blankly into screens, as they attempt to distance themselves from the harsh realities of the mechanized and systemic dominion that they depend on and contribute to, while at the same time sacrificing themselves

> upon the altar of the "green economy." With biomass power plants, solar, wind and hydro farms—all highly dependent on the same ecocidal methods used in their production, operation and maintenance, such as the extraction of rare minerals for circuitry and other sophisticated technologies and the burning of immense amounts of fossil fuels just to keep them in working order and fully integrated into the electrical grid ...[71]

This holistic, or total liberation, understanding of the energy economy, clashed with mainstream environmentalism, which embraced and remained uncritical of low-carbon infrastructures as a way to salvage the industrial economy and capitalism as we know it. The Hambach Forest defenders were pushing for a deeper conversation and realization that challenged relationships of domination and capitalist extractivism whether the energy source was either "green" or "black."

The Hambach Forest employed a diversity of tactics. This entailed an important balance between the Forest and the Meadow. The Meadow provided an important outlet, to be social, organize events, and provide people an outlet to learn, take initiative, and engage in the Hambach struggle, while the Forest could be a more intimate and chaotic space. The anti-authoritarian way the Forest and the Meadow combined permitted a wide array of actions, which included the continuous barricading of roads, tree-spiking,[72] sabotaging, and setting fire to water pumping stations, electrical boxes, and radio masts, which land defenders claim were "being set aflame daily."[73] This extended to burning railway signal boxes, sabotaging the railway, and short-circuiting the electrical rail wires. Furthermore, RWE security personnel were attacked with rocks, sticks, fireworks, and Molotov cocktails, which extended to night actions vandalizing and burning RWE, police, and other company vehicles. The amount of sabotage and arson is nearly incalculable. While the mainstream media, and academics, ignore this, these actions are documented in local newspapers, anarchist news sites, and RWE social media pages. Yet, many more acts remained undocumented and unclaimed. Sabotage actions against the Hambach Forest mine were systemic between 2014–2018. On 31 December 2015, an anonymous communique explained their actions:

> We put homemade harrows (wood planks pierced by large nails) on the road used by the security forces in order to harass and distract them, while we set fire to various blocks of cables and some electrical boxes installed next to the railroad tracks that are used to transport lignite from the mine to the nearby power plants. Stopping the convoys for a while. We then set fire to a telecommunications mast located at the edge of the mine and made sure from a distance that the whole thing went up in smoke. This continued to burn for over an hour. And finally, just after midnight, we attacked again, putting more harrows in the way of the security forces, this time closer to their compound. Then we set fire to a barricade of cars and a large pile of logs on the side of the same road in order to lure the security agents into an ambush, before retreating into the forest with fireworks.[74]

An elaborate game of cat and mouse, sabotage and arson were carried out in defense of the forest, employing harrows as a precaution and trap when destroying mining logistical and support equipment.

Demonstrating a similar approach, another anonymous communique on 26 November 2016 explains:

> After exploring the area, we split up and set fire to six pumping stations, two transformers, an excavator and one of the power grid distribution stations. Pumping stations are key points in the mine's infrastructure and are used to lower the water table and prevent flooding of the mine. Most of the time, they consist of an open-air pipe and an electrical box surrounded by a construction fence. We opened the fence with a crowbar and laid down two simple delaying incendiary devices, along with a pile of bicycle inner tubes, to ensure that the flames would propagate properly.[75]

This, again, demonstrates the multiple sabotage and arson actions taken against the mine, using easily reproducible materials, such as recycling bike tires to damage mining infrastructure.

And finally, this included solidary actions, such as this one in Berlin on 11 October 2018. The report on the attack reads:

> [T]he entrance to the offices of the company RWE-Innogy is set on fire: several incendiary devices are dropped in the front of the door of the mine manager, seriously damaging the entrance at 11 Gaußstraße in the district of Charlottenburg. The attack was claimed, and we reproduce part of the press release: "Despite the temporary stop of the clearing, this is no time for celebration. The destructive machinations of RWE and other energy companies continue undisturbed in other places. RWE provides the fuel that drives global capitalism, whose existence is based on the exploitation, control and devastation of people and nature."[76]

This attack references the temporary hold the courts placed on the cutting season in Fall 2018. In August 2018, BUND filed another suit against the operating plan of RWE in the Higher Administrative Court of Northrhine-Westphalia in Munster, which then in October ordered the stop of the forest cutting.[77] This led to another eviction of the forest. The official reasoning given for the eviction was based on the violation of fire regulations. That eviction was eventually declared illegal by the German courts. The reason why, in the words of Andrea, is because it was "so blatantly rubbish, because they claimed it was for fire protection but it was very much just to allow RWE to continue that crucial cutting season." The eviction took days, resulted in the death of Steffen Meyn,[78] massive protests numbering in the range of 50,000 people, and solidarity actions—like the one above—spread across Germany and the world. Between protests, street actions, death and legal struggles, the Higher Administrative Court of Northrhine-Westphalia ruled a temporary moratorium or stop to tree cutting. This eventually became a permanent decision related to the European Fauna-Flora-Habitat Directive (FFH). While a temporary victory emerges, it would be unwise, as the communique above suggests, to think this mandate to stop the expansive migrating mine is over. These decisions could change at any moment, and there are other coalmines in this region, which are now new sites of protest and the Hambach Forest encampment persists.

THE ANTI-COAL MINING STRUGGLE CONTINUES

The struggle continues around the Hambach mine, and there is continued resistance, not only against the other coal mines in the region, but against lifestyles, policies, and institutional structures that maintain this trajectory of ecocide, qualitative extinction, and irreparable damage to the planet. While a presence continues in Hambach, for the last two-and-a-half years, people have turned their attention to the Garzweiler mine, within biking distance from the Hambach Forest. As mentioned above, this took the form of re-inhabiting the Lützerath village to fight the mine. This, like in the Hambach, entailed creating communal infrastructure (e.g., dry toilets, communal kitchens, and sleeping spaces), organizing festivals, camps, and public actions, all the while evading and suppressing attempts at police eviction and occupation and preparing for the "big day"—when the police finally move in with the full force of the state to evict Lützerath. This day came in January, which resulted in week-long clashes with the police, evading, tearing down their fences and accessing the mine to lock-on, vandalize or burn it. Over 150 people were injured but despite this, people have persisted in quantity and quality of actions, and solidarity and are staying together to fight the ecocidal project of coal extraction in the Rhineland. Lützerath, unsurprisingly, was sold out by green politicians—who negotiated to accelerate coal extraction in an effort to phase it out earlier.[79] This offers an important lesson to the masses of the climate justice movement, and highlights a theme that runs through this book: the state and politicians cannot be believed to stop capitalism and its ecocidal trajectory. And, while Lützerath was lost, the education and practice it provided is priceless for developing an ecological movement with teeth. People are learning, and—as Andrea informed us above— "there has been no 'distancing' from actions and forms of protest over the past week, as often occurs." Civil disobedience and combative militancy, once united, will be far more dangerous and difficult to defeat—people should remain united in their focus on stopping extractive projects, meanwhile developing alternatives to degrow their material and energy use and re-grow community, convivial practices and find what really makes life fulfilling and meaningful in a less destructive way. While it may be easier said than done, if these lessons can be learned, and embodied alongside combative, stress management and healing capabilities, then although Lützerath is lost, the people could be on the road to winning future struggles.

These experiential lessons, however, must be used to combat the next round of political manipulations and maneuvers. Green capitalism, as eco-anarchists in the Hambach remind us, is not a solution to the ecological crisis but instead old wine repackaged in new bottles. Bottles are designed to conceal the mineral mining, chemical leaching, smelting, manufacturing, shipment, and decommissioning of wind, solar, and other lower-carbon infrastructures. Lower-carbon infrastructures, electric vehicles, and digitalization schemes remain the "new hope" of financial capitalism, which through modeling gymnastics have managed to brand themselves "green" and "environmentally friendly" (see Chapter 1). People tend to underestimate the political, human, and electrical resources that go into producing the material necessary to be "green," "clean," "sustainable," and "renewable." Wind turbines, high-tension power lines and literally every electrically powered infrastructure and gadget, are going to require a plethora of minerals. This remains the main focus for the remainder of the book, which in the next chapter, moves to southwest Peru to look into the reality of copper mining, We too learn from the communities in the Valle de Tambo, whose struggle and resistance teach us the true cost of the copper that powers your cities and electronic devices. And, we must remember that copper is only one of the many mineral and hydrocarbon ingredients that compose these infrastructures and electronic devices.

4

Mineral demand: the Tambo Valley struggle against copper extraction and state terrorism

While traveling from Arequipa, Peru, to the Tambo Valley (Valle de Tambo), two hours southwest of Arequipa, we eventually came out of the mountains and into an unexpected desert plateau. Toward the end of the journey, sitting in a *colectivo*—collective passenger van—with my friend Carlos, who invited me to the region, we stared out the window at a sandy desert landscape. Intense, as it felt more desert or arid than I was used to in California and Arizona (e.g., Shoshone, Hopie, and Navajo lands). The desert plateau above the Tambo Valley was vast, and was demarcated with signs along the highways alerting travelers of a military zone, used by the Peruvian air force for bombing practice and, also surprisingly, for Mars space expedition tests, as this area has conditions that resembles the planet Mars.[1] Considering I was traveling to an agricultural valley, fighting to defend their land, river, and agricultural practices against a Mexican mining company—Grupo Mexico's subsidiary Southern Copper Peru—I was shocked by how arid the region was just 15–20 minutes outside the Tambo Valley. As we drove into the Valley at dusk from the desert plateau, the region appeared like an oasis. Desert turned into green fields, trees and a glistening river at the center of the Valley surrounded by little cities, scatters of small houses—some made of cinder blocks—and other habitations that looked a bit like shacks or favelas on the surrounding hills. When I turned to Carlos to express my surprise at the geographical contrast, I said: "So the mining company wants to take all of this?" Pointing to the river, the vibrant green fields, and trees. He replied: "Si" (yes).

The Tía Maria conflict began in 2009, yet Southern Copper Peru, hereafter referred to as Southern, already began assessing the mineral reserve situated above the agricultural Tambo Valley in 2000. It negotiated with government officials and civil servants in 2005 and later provided three consultations (*audencias*) to the Tambo Valley. The third consultation in August 2009, however, is when open conflict broke out. In the previous two consultations, there had been extensive debate over how Southern would access and use water in the region. The company initially said it would create a desalination plant and pump water in from the sea, which was approximately 20 km away as a bird would fly from the mining site. In the third consultation, Southern changed their position. Presumably to avoid the enormous costs of desalinization and pumping in water at a greater distance, Southern announced its preference to use the ground and river water—and not sea water at the mine.[2] This triggered outrage. People began rioting, throwing projectiles and plastic chairs at Southern representatives after this announcement. What began here, would develop into a protracted conflict with strikes, barricades (see Figure 4.1), and self-defense against invading police forces that, since 2011, has resulted in eight

Figure 4.1 Protesters stand behind a barricade in Cocachacra. Photo: Miguel Mejía Castro.

deaths—seven protesters and one police officer—hundreds of injuries and President Ollanta Humala declaring a 60-day State of Emergency on 9 May 2015. Despite political turmoil and instability, the Tía Maria mine has not begun operations and the struggle continues to this day. This struggle, however, has been brutal and extensive, inextricable from the violent political history of Peru that has seen challenging conditions for anti-extractivist struggles and societies-in-movement.

CONTEXTUALIZING PERUVIAN EXTRACTIVE POLITICS

Between the early 1980s and 2000, Peru was engulfed in a civil war or, more accurately, a "dirty war" between two authoritarian actors. The Shining Path (*Sendero Luminoso*), a Maoist political party that arose from conditions of capitalist society cultivating extreme material poverty, declared an insurrectionary war in May 1980.[3] While the government initially did not take the Shining Path seriously, this Maoist group's ability to organize a popular base and execute military action eventually struck fear into the Fernando Belaúnde administration. It responded by declaring a State of Emergency and, in the words of Steve J. Stern, "provided the military a platform to conduct an Argentine-style 'dirty war' against presumed subversives,"[4] resulting in the torture, murder, and massacre of people—journalists, farmers, Indigenous peoples—in the Ayacucho and other highland regions in Peru. The Shining Path, however, engaged in similar terrorism style tactics and became well-known for executing anyone opposing them—notably Indigenous leaders and farmers.[5] The extreme violence and terrorism enacted by the Shining Path would eventually weaken their support, creating openings for the state in the form of state-sanctions, but also autonomous militias, known as *rondas campesinas* (peasant patrols) that organized self-defense networks against the Shining Path.[6] "The anti-authoritarian thought memory of the left of the 1970s and 1980s," Peruvian anarchists explain, "was lost as a result of the persecution they suffered under the dictator Fujimori as well as the genocidal Shining Path—dogmatic communist who murdered campesinos, leftist leaders, and anyone else who opposed them."[7]

Before almost toppling the Peruvian government, the Shining Path was eventually defeated in the mid-1990s, but this "dirty war" had a lasting imprint on Peruvian politics. While the rippling effects of it are countless, two points are of special interest in relation to the Tía Maria mine. First, is the discourse of terruqueando, emerging under the Fernando Belaúnde administration in the early 1980s, which criminalized any challenge to the existing political order as "subversive discourses"—lumping all politics together that expressed disastification against his regime.[8] This, of course, was designed to eliminate vocal support for the Shining Path, and other revolutionary groups. Terruqueando, or terruco (terrorist), would gain cultural significance when President Fujimori would use this to describe anyone questioning his neoliberal policies or political rule.[9] While this relates to the general trend of governments calling anything that challenges their policies terrorism,[10] terruco has racist connotations and "is the adjective used to define who can be killed with impunity."[11] Recently, during the Defend Atlanta Forest Struggle, which seeks to protect a forest ecosystem from a police urban warfare training center in the US,[12] police not only killed Manuel Esteban Paez Terán, or "Tortuguita," for sitting in a tree house (to prevent the tree from being killed), but the subsequent protesters were also charged with terrorism.[13] This tactic of governments branding people resisting infrastructure and mining projects as terrorists, indicates an expression of totalitarianism, which, in Peru, has a particularly violent and racist history that reinforces extractive security policies.

Second, is the collaboration between extractive corporations and Peruvian armed forces. In the 1980s, the Shining Path's weapon of choice was dynamite stolen from mines, later used to target civil and extractive infrastructure.[14] The deployment of terrorist tactics by Shining Path served as a pretext for a series of laws: the Defense System Law (1987), the Organic Law of the Defense Ministry (1987), and a series of decrees in 1991, 2002, and 2007, allowing the military not only to "assume control of internal order during States of Emergency,"[15] but also to contract with private companies. In *Military Politics and Democracy in the Andes*, Maiah Jaskoski[16] reveals the military's aversion to engage in domestic counterinsurgency operations, which made the Peruvian National Police

(PNP) the preferred practitioners. This relates to policy changes under the García administration in the late 1980s, where it was declared military personnel could be held accountable for their crimes committed during military service.[17] As a response, in the 1990s and early 2000s, "the national police increased its participation in counterinsurgency considerably" against Sendero forces, explains Jaskoski,[18] to the point that "the national police force has expanded its counterinsurgency activities, potentially encroaching on the army's domain and threatening its future budget share." While the PNP became the leading specialist in counterinsurgency operations, the military's relationship with extractive companies grew. Moreover, the constitution encouraged, under "resources directly collected" (RDR) contracts, the military to rent equipment (vehicles, helicopters, and specialized gear, etc.), infrastructure, and personnel to private oil and mineral companies. Contracts were drafted with local army commanders, and between the years 2003–2005 RDR comprised approximately 8 to 11% of the defense budget and could cover between 5 and 50% of military base operating expenses. This work proved highly lucrative for military commanders and offered benefits for soldiers, such as medicine, discounted airfares, and access to more desirable food and lodging.[19]

Extraction companies became increasingly influential in directing military patrols, hiring "Peruvian army units to conduct counterinsurgency patrols" and, in some cases, the army "conducted police work for companies."[20] Notable among these companies, according to "a former private security official," was Southern Copper Peru,[21] which "received army protections, suggesting that private mining companies have proved exceptionally influential in terms of affecting army behavior." The "securitization of strategic resources" then, as highlighted by Middeldorp and colleagues, becomes a reciprocal self-reinforcing process: "the military is deployed to guard the extraction in progress, and resource extraction revenues are in turn invested in the military."[22] This self-reinforcing cycle is not limited to the extractives sector, and is more broadly implicated in the increasing militarization and marketization of nature that I mentioned previously.

Under President Alan García (2006–2011) private–public security partnerships would expand to include the Peruvian National Police. The construction of Indigenous land defenders as the "internal enemy,"[23] once reserved for the Shining Path paved the way for the 11 July 2009 Decree that authorized "the provision of extraordinary additional services by the police,"[24] in the form of two types of contracts: institutional and individualized. Institutional contracts require an agreement between the Director-General of the PNP and the persons or entity requesting protection, which can classify as either permanent or occasional services. Permanent service would be for a specific time period and occasional would be for short requests of between one to eight hours. The individualized "extraordinary additional services" are performed by off-duty police officers, which only require an agreement with the individual officers[25]—the latter could be read as legalizing paramilitarism. The PNP are currently servicing over 22 mines in over eleven regions with more than 485 police officers working with mining companies "to 'prevent, detect and neutralise' threats by means of precautionary measures, surveillance and patrols."[26] Threats are understood as "criminal actions, assaults, acts of sabotage, and terrorism," which could also include "acts of a threat-like nature" specified as "civil war, invasion, insurgency, strikes, internal unrest, civil disturbances, rebellion, vandalism and other criminal and terrorist action."[27] Jaskoski demonstrates that the police—specifically the PNP—have out-performed the military and have taken the lead in counterinsurgency operations domestically.[28] Notable among the PNP is the DINOES (National Division of Special Operations) trained for "anti-subversive" activities and deployed to support extractive operations.[29]

Expanding counterinsurgency by Peruvian security agencies dovetails with plans from the United States to send over 3,000 US soldiers to Peru for narcotics interdiction and to combat insurgents, which School of the Americas (SOA) Watch[30] believes is "a guise for military control and repression of social movements, especially those defending their natural resources."[31] This proliferation of private–public security agreements has transformed the public security forces into private contractors. Furthermore, according to the January 2014 Law No. 30151 members of the armed forces and PNP are exempted from criminal responsibility if they cause injury or death on duty. Human rights groups have called Law No. 30151 a "license to kill."[32] These arrangements are also used by Southern Copper, which registered an agreement in 2010 with PNP XI Dirtepol of Arequipa to "provide extraordinary services" under an "individualized service" contract.[33] Thus impunity is granted under Law No. 30151 to security personnel, which appears to have taken effect during the Tía Maria conflict.

The impact of military, police, and paramilitary services to suppress anti-mining struggles is alarming. The impact of these policies has enormous environmental implications. For example, the Peruvian state's Defensoría del Pueblo registered 224 social conflicts in 2013, 149 (67%) were socio-environmental conflicts and 108 were related to mining activities.[34] By 2014, mining concessions occupied 20.42% of the country (Romero 2017: 15), while mining investment in Peru increase 3.8%, totaling USD 2.833 billion in August 2017.[35] Organizing the country around an extractivist model of development, an economic growth strategy centered on market-based natural resource demand and extraction, Peru has been littered with various extraction projects and, consequently, a plethora of environmental conflicts with rural and Indigenous populations.[36] This extractivist development model, and resulting conflicts, has made the Peruvian military and police, especially the Peruvian National Police, companions to extraction companies.

THE TÍA MARIA PROJECT AND CONFLICT

Southern began conducting extensive geological and geochemical studies in 2003, which was followed by the Ministry of Energy and Mines (MEM) granting approval for an Environmental Impact Assessment (EIA) in 2006.[37] Since 2012, however, the MEM has become responsible for approving EIAs and not the Ministry of the Environment.[38] The Tía Maria project seeks to extract 120,000 tons of copper cathodes per year for 18 years with a 1.4 billion-dollar investment, which would include three mining and processing sites. The first mining site is "La Tapada" in the Pampa Yamayo, located closest to Cocachacra, El Fiscal, and the Tambo River. Exemplifying, Stuart Kirsch's notion of "corporate science,"[39] Southern measurements claim that La Tapada is 3 km[40] away, while independent investigators demonstrate it is 1.2 km[41] and locals assert that the distance from the Tambo River is between 500 and 700 m.[42] Second is the "Tía Maria" site in the Cachuyo area, which according to the company is 7 km from the Tambo Valley and, third, the processing and leaching site in the Pampa Cachendo that is 11 km away from the Valley. Tía Maria is an extensive mining operation, which will also produce more than copper—even if other highly valuable minerals, such as gold, are rarely talked about.

Southern entered the Tambo Valley in a similar way to wind energy projects in Oaxaca discussed in Chapter 2 by approaching regional and local authorities. The company began by approaching national political bodies, local municipal leaders, and, eventually, civil society groups. President of the Broad Front of Defense and Development Interests in the Islay Province, at the time, Catalina Torocahua, explained that in "2006 the mine became known as a result of usurping city boundaries" and by "2007 the company entered formally to talk with the authorities: Mayors and leaders."[43] At the time Catalina was trying to start a portable water project for the Valley, called Plan Maestro. Catalina recounts how one representative from Southern approached her, explaining that if she accepted the mine her portable water plan "will be achieved, because the mine is going to give a big canon minero, and supposedly this canon minero will make this project a reality." The 1992 canon minero law, Anthony Bebbington explains, "allows 20% of corporate mining tax to be allocated to the territories where companies operate. In 2001 the canon minero was raised to 50% and extended this tax to other extractive activities."[44] Catalina, aware of the ecological costs of mining, expressed profound hesitation based on her knowledge of nearby mines—between one and three hours away—in Moquegua, Tacna, and La Oroya. The representative replied by assuring her that there was an "abundance of water," that pollution and particulates from the mine would not spread throughout the Valley, covering people and their crops. Catalina responded by asserting that "What is logical is that no mayor or leader decides, it should be a *consulta popular* (popular consultation) and nobody should oppose this." According to Catalina,

Southern's representative replied: "I have already talked to the mayors and leaders and they already agreed." This response generated dismay and frustration among residents.

Opposition to the Tía Maria project was based on experiences with Southern, and their conduct, but also with mining in general. Catalina and other research participants referenced a series of mines: La Oroya, a distant copper smelter; Moquegua[45] and Toquepala[46] copper, molybdenum, rhenium, and silver mines operated by Southern in the neighboring province. People also referred to the Tacna mine operated by Minsur; the Cerro Verde copper mine; and, most importantly, Southern's smelting facility in Ilo. People claim that Ilo produced acid rain in the Tambo Valley, killing all the olive trees in the 1970s, which is a common story linked to the belief that Southern authorized the assassination of agronomist and opponent of the Ilo smelter, Carlos Guillén Carrerra on 2 October 1998.[47] In short, Southern already had a contentious reputation for their operations and political activity.

Previous experiences with mining projects combined with the Tambo Valley's strong agrarian culture laid the foundations for ardent resistance. There are about 40,000 people, José Romero's contends in his new book—*Lo Que Los Ojos No Ven* (*What the Eyes Do Not See*)—that are anchored directly or indirectly into the agrarian economy.[48] Cocachacra, according to Romero, consists of 2,000 plantation owners, 7,000 small holders and renters along with 8,000 day laborers (*Jornaleros*).[49] The district of Cocachacra, based on 2007 census data, consists of 47.15% agricultural activity, 11.72% in retail trade, 5.63% in transportation, 4.36% in hotels, 4.16% in construction, and 3.53% in mining and quarries.[50] While most people engaging in agriculture also have small commercial operations, the Cocachacra demographic is similar in other districts, often with agriculture dominating the region.[51] The Tambo Valley retains a strong agrarian economy, making agricultural works the backbone of people's livelihoods and existence in the area. The mine, in short, is rightly understood by the vast majority of residents as threatening that livelihood and heritage.

There are three principle reasons why the general population rejected the mine: (1) ground water usage; (2) contamination of the ground water, and (3) air pollution. Southern claims that the wind blows on shore, implying that the wind will not carry mining particulates into the Valley, yet as Catalina and others pointed out the wind's current shifts daily between on and off shore, highlighting a popular misconception propagated by Southern, that further entrenches existing distrust between the local population and the company. Regarding socioecological impact, it was common for residents to feel that "the mine is only destruction"; "it is slow death," and, referring to social development funds: "What is the point of having schools, medical clinics, universities if they will be slowly killing us?"[52] The link between slow violence and mining is clear to the majority of residents.[53] Furthermore, the job opportunities offered by Southern are not only limited compared to those offered by agriculture in the Valley, but they require technical expertise that will discriminate against the young, old and unskilled labors. Romero contends that Tía Maria's "operation phase would hardly employ 600 workers, while the agrarian economy offers jobs to more than 20,000 families."[54] Lastly, like other research participants, Catalina rejects the idea of authorities approving the Tía Maria mine without achieving the population's general consent or "social license."

There were three public consultations hosted by Southern in November 2007, July 2009, and August 2009. Despite popular skepticism and distrust rooted in experiences with Ilo, Moquegua, and disinformation about the wind currents, the consultations gained steam and the company offered three methods of mine water use, from the river, ground or sea water. During the August 2009 consultation, the EIA did not favor desalination of sea water and residents—based on identity cards they found—claimed that Southern bused people into the consultation "from the outside," notably students from the National University of San Augustine in Arequipa to fabricate consent for the mine.[55]

Consultations as spaces for manufacturing consent are not uncommon,[56] yet this triggered a riot and assaults against representatives of Southern with rocks, sticks, and plastic chairs.[57] Now the Defense Front and Interests of the Tambo Valley inspired by previous struggles in Tambogrand and Minera Majaz,[58] began to implement Catalina's idea to hold a popular consultation (*consulta popular*) against Tía Maria and its EIA. In October 2009 the

Defense Front organized the popular consultation in Cocachacra, Punta de Bombón, and Deán Valdivia, resulting in 93.4% rejection of the Tía Maria project by the voters.[59] This led to locals continuing to organize protests and strikes, called against Southern in 2010, while 4,000 police (PNP) were ready to enter the Valley, the MEM opted to contract the United Nations Office of Project Services (UNOPS) to evaluate the Tía Maria EIA.[60] Then in March, just before UNOPS would release the report, the contract was canceled, due to budgetary constraints by MEM.[61] The UNOPS report, however, was leaked. The leak noted 138 observations including missing a hydrological study and failure to recognize other minerals (e.g., molybdenum, silver, gold) in the concession.[62] When it became clear that the government would ignore UNOPS' 138 observations an indefinite strike was organized for 23 March 2011.

The state responded with repression. The PNP flooded the area, police attacked and protests escalated, which gave rise to the now infamous Espartambos (Figure 4.2). Referencing the Spartans from the film *300* in the Tambo Valley, these individuals took a position of combative self-defense against the police, widely recognized for

Figure 4.2 Espartambos on the frontlines battling police and their armored vehicles, 2015. Photo: Miguel Mejía Castro.

grabbing large sheets of tin and wood to use them as shields to block the rubber bullets, bird shot, and rocks of the police. The conflict escalated, with the police shooting and killing four: Adrés Taipe Chuquipuma on 4 April followed by Néstor Cerezo Patana, Aurelio Huarcapuma Clemente, and Miguel Ángel Pino on 7 April 2011.[63] This led the government to temporarily pull back from the Tía Maria project as the strike paralyzed the Valley and the conflict sowed terror and resentment.

This victory, however, was short-lived. Southern launched an information campaign to re-enter the Valley. According to Carlos Aranda, Southern's national head of "community relations," in late 2012 were "summoned by the Ministry of Energy and Mines" and "they told us that: 'we are willing to give it another chance, so let's go ahead with Tía Maria.'"[64] The government imposed three conditions on Southern: (1) a desalination plant; (2) redo the social impact study; and (3) comply with the UNOPS' observations. Aranda, explains: "We did all three of them."[65] In January 2013, Southern began a full-spectrum public relations campaign, and it began a new EIA in November 2013. Meanwhile, people commenced hunger strikes against Tía Maria in Arequipa in December and October 2013.[66] Eventually, however, the new EIA was approved on August 2014, as MEM claimed that all the observations were resolved. Despite the deaths, investment risk, and popular opposition, the government decided to permit the project.

The Mining Conflict Observatory (OCM), however, found the same issues as with the previous EIA regarding the hydrogeological and subsoil analysis, and the lack of information about the crushing plant and tailing management. Other objections related to particulates from construction, and the risk of sulfuric acid evaporation causing acid rain. Moreover, the new study evaded assessments on the desalination plant's impacts on nearby wetland conservation,[67] and also neglected community participation in the study.[68] These irregularities led the Defense Front to request UNOPS to review the new EIA, which the government denied.

Once negotiations broke down an indefinite strike was called on 23 March 2015. Again, the Valley erupted in protests, road blockades, and eventually combat with the PNP (see Figure 4.2). Then Southern's Tía Maria head of "community relations," Julio Morriberón, declared on 27 March once again, the retirement of the Tía Maria project, first due to, "the onslaught of a new type of terrorism, anti-mining terrorism" and "second, the paralysis of the state in its role of promoting investments and giving the necessary guarantees to get them started."[69] Tapping into the socio-politico discourse of terruco, or terrorist, Southern employs a cultural device from the "dirty war" that constructed (dark-skinned) rural and Indigenous people as subhuman, fanatical, and violent terrorists,[70] thus criminalizing and, in the past, justifying scorched-earth counterinsurgency tactics against highland populations during the war. This trend has continued since,[71] and recently after Primitivo Evanán's exhibition, at The Art Museum of Lima (MALI), was slandered as "terrorism apologetics" (*apología al terrorismo*); journalist Gabriela Wiener,[72] highlighting this socio-political device, coined the verb "terruquear" to refer to a "political strategy that uses the fear of terrorism for its benefit."[73] Morriberón, and, later, Aranda,[74] we can say are terruqueadores, people who are operationalizing the discourse of terrorism from the "dirty war" to encourage, and subsequently, justify, police and military intervention to break strike, blockades, and enforce the operation of the Tía Maria mine.

In April, the conflict further escalated. Three residents were murdered by security forces: Victoriano Huayna Nina, Henery Checlla Chura, and Ramón Colque, and one police officer, Alberto Vásquez, was killed. This eventually gave way to President Ollanta Humala declaring a State of Emergency on 9 May 2015, involving the presence of 3,000 police and 2,000 military personnel.[75]

STATE TERRORISM FOR COPPER

The contested approval of the EIA and the deployment of the PNP and, later, the military against the strike only reinforced the existing belief that the politicians nationally and locally had been bought by Southern. "The politicians just want to fill their pockets because they do not care about the consequences, the consequences that will affect everyone—they are not going to live here," explains a disgruntled mother. "They promise us [to terminate Tía Maria] every time they come here, but when they enter as presidents they begin to agree with the contracts and all of this—they are cheats."[76]

President Humala came to the Tambo Valley before the election in 2011 proclaiming that Tía Maria "must be revoked" and that the people's "voice has a binding character before any political decision,"[77] then four years later declared a State of Emergency against the Valley asking that the "full weight of the law come down on these criminals, murderers and extortionists."[78] Local politicians and administrators tend to have a more antagonistic dynamic with the mine, but, as Romero reveals, there are "mining candidates" sponsored by Southern "with the aim of legitimizing and promoting their interests."[79] In December 2017, Yamila Osario, the current governor of Arequipa, accepted a 770,000 soles (approx. USD 235,503) donation a week before announcing Tía Maria will begin operations in March 2018.[80] The Odebrecht bribery scandal, which placed ex-President Humala and his wife in pre-trial detention and caused President Pedro Pablo Kuczynski to resign, serves as a high-profile reminder of the systemic political corruption, which residents associate with the Tía Maria project.[81] Distrust and contempt for politicians would characterize the majority of interviews, but, in a neoliberal regime, the line between bribery and contract negotiations, the public and transnational private sector is increasingly fine.

The Tambo Valley residents are convinced that the PNP and military work for Southern. "In reality it is not the state that declared the State of Emergency, it's the project that induced the state through the power they have to declare a State of Emergency, for what? Why? ... to minimize protests," says a conservationist, who continues: "there are cities with crime, theft and a lot of bad things, but there are no police," but "there are police to defend a project, a hill, but there are no police to defend and take care of the population—that is strange, no?"[82] Southern's Carlos Aranda, however, describes the situation in 2011 this way:

> We had three people that were killed in this process. Not because they were on our property or in part of the project, they were killed because the mob decided to block the roads and in Peru that is illegal. When they went to block a major highway, like Pan American South, and they blocked it for days, the police came in and violence took over the Valley and three people died.[83]

Southern representatives stress that Southern adheres to "environmental regulations" and the "rule of law," while the protesters do not. Later with cheerful condescension, Tía Maria's head of community relations explains: What "happened is regrettable, your little deaths (*muertitos*), rest in peace, who died in these protests," continuing to assert that "it's true that Southern, like every human being, also has the capacity to say, 'stop.'"[84] Aside from the poor choice of words, "muertitos," this statement defends the transnational corporation's right to stop popular oppositions to their operations.

The 2015 indefinite strike unleashed the fury of the state (see Figures 4.1, 4.2, and 4.3). It began with marches and road blockades (see Figure 4.1), with women, children, and the elderly leading the demonstrations, when, according to accounts, "the police came and pushed, used their batons, and then the people also used sugarcane sticks and the police began to use teargas bombs. Then the people responded with whatever they had." In 2011 and 2015 police actions were brutal: beating people; shooting tear gas, slinging rocks from Hondas—long-range slingshots that you swing over your head—and even firing bullets at people. Undercover police and informants were deployed; the area was monitored with drones; and the air was patrolled with helicopters that also fired teargas and dropped rocks on the demonstrators. The PNP further engaged in acts designed to create depredation and psychological stress by burning rice fields; and even attacking funeral processions with teargas for people killed during the demonstrations; as well as the repeated accounts of police framing people once they were arrested. A well-known instance is the case of Antonio "Miguelito" Coasaca. On 23 April 2015 the PNP DINOES division beat and dragged Coasaca down the highway, then concussed, the police broke his hand and forced a caltrop—an anti-tire road spike—into his hand. Subsequently, they called a nearby journalist over to take a photograph as evidence. This police action was documented on video[85] and later charges against Coasaca were dropped in court.[86] The lawyer Héctor Herrera, who defended Coasaca and over 120 people from the Valley during this period, remembers: "I also have people who were planted with bullets, dynamite, and every day somebody arrives to me with rocks, sticks, and Hondas and the police saying: 'I saw him with that.'" Herrera continues, "here is a delinquent state who ordered the police to plant evidence and prosecutors that collaborated with them."[87] The police also covered up evidence by taking bodies from hospitals, which would additionally delay and prevent medical care. "[N]ot only did they kill us," explains Kali, "but when we wanted to take the body of our brother they threw tear gas canisters and they prevented us from taking our dead."[88] Ramón Colque "bled to death" and "a lot of people wanted to help him, but 'the police did not allow us,'" explains a woman continuing that the police "stole the body from the clinic and I think they wanted to disappear him or something like that."[89] Scuffles over corpses and the wounded in clinics and hospitals were frequent during the indefinite strikes.

Figure 4.3 A protester flings rocks with his Honda in the hills above Cocachacra. Photo: Miguel Mejía Castro.

The violent behavior of the police only escalated during the State of Emergency. Police cut off the lights in the Valley, threw rocks and teargas into houses, and conducted night raids on houses of suspected organizers or Espartambos. Two women explain:

W1: It was a terrible abuse, terrible abuse, and they did whatever they wanted with us. They entered into the houses at four in the morning to take out all the young people, because snitches were whispering.

W2: By pointing people out ... they entered into Maria's house in the middle of the night to take out her husband.

Q: With what reason?

W2: They said that he was an Espartambo.

Later in the conversation this second woman recounts the police telling her to "shut up you shitty terrorist," when she was pleading with them not to throw tear gas into her house where her sick mother with bronchitis and "a delicate heart" was living. The police did anyway and they had to get her to the hospital in Arequipa.[90] The trauma left by the police would surface with eyes of rage, fear, and tears. During interviews older women would repeatedly break into tears remembering how the police chased, beat, and dragged people down the street or when the police would raid houses in the early morning dragging people out of bed, while children stood naked in the street. Meanwhile, the military stood by holding assault rifles watching, and by some accounts even criticized the behavior of the police. One person recounted that one military officer even said, "Sometimes I want to grab my weapon and shoot the police. Because I do not like how they treat the old women—they push and kick them"[91] Not only does this resonate with earlier accounts by Jaskoski, that the Peruvian military refrains from domestic operations,[92] but also Hanna Arendt's early observations of the Nazi regime and the enhanced abilities of police to dispense and orchestrate terror.[93]

Because of the brutal behavior and murders by the police, along with them having "long hair, scars on their face, beards, [and] tattoos,"[94] people were convinced that these were not actually police officers, but rather mine security personnel—"they weren't police they were miners dressed like police, yes, contracted by Southern."[95] An ex-military private security contractor, working for extractive companies for over 15 years, let's call him "Jim," explains that in "a part of the intelligence service here in Peru, exist the famous mercenaries" that are sent to the "frontlines" of conflicts, "because when the police commit excesses, they get in trouble." According to Jim, the police contract mercenaries for purposes of plausible deniability, which complements accounts in 2015 that the PNP had unknown name tags that read "FilosofeXXX."[96] When asking Jim about this specific instance in the Valley he replied: "In some cases, they dress by themselves like police, because they also want to protect themselves." While Jim only confirms these tactics as common, they overlapped with the deployment of undercover police and informants in the demonstrations. Many land defenders caught informants, who would confess that they were paid 100 soles to attend meetings and demonstrations.[97] Isabel explains: "When people captured infiltrators, more than three or four people, also women, they would punish them because they were paid by the mine to infiltrate, but we already knew who they were. When we see people we do not know, we already know it's them."[98]

In a strike and blockade (*el paro*) against an Ecuadorian oil contractor, Japhy Wilson offers a detailed, and gripping, account of how that movement would identify, and eject company and police infiltrators, which, similar to the Tambo Valley, relies on the existing relationships and social fabrics to identify and kick out police spies from protests.[99] Wilson's account also reveals the pitfalls and challenges of committed researchers entering struggles from the outside to avoid being branded as infiltrators, and of living and struggling in common with the movement. During the State of Emergency, informants were also used to point out the homes of protest organizers and Espartambos—"So there were police infiltrators and they took notes where people live, and in the night the police would abduct people," explains a local resident.[100] Police infiltration should be expected in any struggle. Important, however, when identifying police spies or informants is for people to stay calm and to begin a process of collecting evidence and managing them to minimize the type of information they can gather from the struggle. Accomplishing these steps will make "outing them" and kicking them out of the movement more effective, as mistakes are sometimes made and it is best to avoid making them. Police or company infiltration is intended to generate paranoia, stress, and confusion, avoiding these traps are central to successful movements. These infiltrations, and other repressive activities, are part of tiring movements, and the "war of attrition" land defenders speak about in France (Chapter 5). It is crucial that land defenders, or protesters, actively avoid self-important paranoid narratives, which means taking mental health issues seriously and developing practices (e.g., therapy, Chinese Medicine, Qi Gong, Yoga, etc.) in order to overcome the fear and paranoia that police and companies are seeking to perpetuate.

The lawyer, Herrera, suggests the possible presence of Grupo Terna, a division of undercover police, were present in the Tía Maria conflict. Jim, however, explained that not only in "the majority of the cases" mining companies "work with DINOES," but that "every mining and oil company has their own [secret] service that we call, special services." This security contractor continued to explain how every resource extraction company in Peru has something called "Internal Affairs" (*asuntos internos*), which is the extraction companies' "intelligence service," largely staffed by ex-military and security personnel, that specializes in "counter-intelligence" and is responsible for neutralizing opposition against the mine and stifling attempts at corporate espionage and concession theft. Jim explained that the job of Internal Affairs was to pacify opposition, which included negotiating "1 million" soles with local opposition leaders to stop protests for a certain amount of time. Narrating what companies call "community relations," Jim continues:

> So they deceive them [locals] with a medical clinic, I do not know, with a donation of clothes for kids, or blankets, so something that looks great for the community, but in reality it is nothing. So they handle it this way. So what does Internal Affairs do? They begin to check up on the leader

and if the leader begins asking for more or breaks his deal by protesting to negotiate for more money with the extraction company, then "sometimes they are in charge of disappearing him from the map." This resonates with the Pepe Julio Gutiérrez case, former Tambo Valley Defense Front leader, who in January 2018 was sentenced to 30 years and six months for negotiating 1.5 million "lentils" to end the indefinite strike with a lawyer from Southern in 2015.[101] Romero contends this was a strategy to fragment the social movement, which remains plausible as Southern's lawyer was not brought to court.[102] State practices in Peru operate along the lines envisioned by General Kitson's recommendations on how to use the law as a method to neutralize activists and "disposal of unwanted members of the public."[103] Nevertheless, the dark side of environmental conflicts is revealed—cooptation, assassination and "neutralization" by every means.

When discussing the political violence in the Valley during 2011 to 2015, people at the demonstrations or families of the dead were convinced that snipers were present at the demonstrations. A woman, for example, exclaims: Néstor "got shot in the head, which is not a lost or accidental bullet. We are not that foolish—we recognized that a sniper was there."[104] Raising this reoccurring issue to Jim about assassinations in the Tía Maria conflict, it elicited this response:

> J: That is what we call asymmetric warfare. So with this they [the company] tell you, "If you continue irritating me, the same is going to happen to you." This is another way to debilitate the group, if you are the principal head [protests leader] and you move 20 people, I am not going to kill him [another person], I want to kill you. Because when I neutralize you, I weaken all of them. And now, who is going to direct them? Nobody is going to direct them, so this group has to create another and I come back and I kill him again.
>
> AD: And for the asymmetric warfare, how involved are the mining companies in strategies of asymmetric warfare?
>
> J: They always do it, they always use it. As I told you they use it to debilitate the group.
>
> AD: Asymmetric warfare is part of what they have to do?
>
> J: It's part of the process.
>
> AD: It's always part of the process?
>
> J: This case is more of a last resort. When the situation gets very difficult ... when a problem has been created that cannot be con-

> trolled. When they cannot control the situation with money, this is the last resort and especially when the leader, for example, receives 1 million [soles], but in two or three months he goes and asks for more money.

This animates where and how the "soft" and then "hard" approaches to managing opposition meet, demonstrating the intricate relationship between counterinsurgency and resource extraction operations. While Jim did not work for Southern, he confirms that there are people "still in the army … working intelligence … with them."[105] It is clear, Jim continues, that "without asymmetric warfare techniques the protesters can paralyze the mining operations, and in one day the company will lose millions." This intimately emphasizes the importance of resistance to stop mines and counterinsurgency to prevent the building of mines, revealing how the police, military, private security, and the doctrine of scientific violence are instrumental in enforcing extractivism.

DEVELOPING THE VALLEY: PUBLIC RELATIONS, SOCIAL DEVELOPMENT, AND VALLEUNIDO

The intensity of police violence was accompanied by Southern's "information campaigns," "community relations" teams and development programs that, one woman echoing inhabitants in the Rhineland said, "entered into everything!"[106] Catalina agrees, "they are in everything, they are in education, they are in health care."[107] Southern's operations have entered into just about every facet of Tambo Valley life, manifesting an extractive biopolitics, rooted in counterinsurgency tactics to secure access to subsoil resources. This socially integrative biopolitical approach is consistent with, and animated by, previous research on "soft" counterinsurgency and corporate social responsibility (CSR) initiatives.[108] Research has linked extraction company social interventions to *The Insurgencies and Counterinsurgencies Field Manual* chapter titled "Indirect Methods for Countering Insurgencies," specifically the approach known as "*integrated monetary shaping operations.*"[109] "Integrated monetary shaping operations are the coordinated use of money, goods, or services to support" the goals of security forces (or mining companies), using "developmental assistance, infrastructure, and governance support projects to win the support of an Indigenous populace and erode support for the adversary."[110] There is a strong relation between US and Peruvian counterinsurgency,[111] which even extends to FM3-24 explicitly referencing Peru as successfully employing integrated monetary shaping operations:

> Peru demonstrates that intelligence capabilities can be integrated with information operations and integrated monetary shaping operations to successfully undermine an insurgency. The Peruvian government was eventually successful in using economic development and an information campaign to weaken the Shining Path insurgency.[112]

Southern's approach in the Tambo Valley closely resembled the principles of integrated monetary shaping operations to "win" the "hearts" and "minds" of the population to mitigate resistance and gain "social license" for the Tía Maria project.

In January 2013, Southern Copper began a full-spectrum "intervention"[113] into the social life of the Tambo Valley. This first began with "Plan Reencuentro" which would invest 100 million soles (approximately USD 30.6 million) into disseminating information about the Tía Maria mine, with 46 volunteers going door-to-door, offering to paint and give new concrete floors to houses.[114] Aranda, explains: "[W]e started doing things, like paint your house, because after the violence a lot of the houses were left in really bad shape with graffiti and you know that sort of thing." House painting, especially after 2015, served a dual purpose of community relations, but also to paint over the anti-Southern graffiti in the region. Social development, mining public relations and "Broken Windows" theory—the theory that visible symbols of civil disorder, or political opposition, encourage further crime[115]—originates into confluence in this counterinsurgency strategy.

People in the Valley, however, saw what was happening and, in Aranda's words, "some of them [staff] had been kicked out of the Valley. They ran away, there were threats made to their lives, people tried to burn their houses and [vehicles,] so they left the Valley." This terminated Plan Reencuentro, which was later revived as "Future Arrived" (El Futuro Llegó) in 2015 while the second EIA was under review. This initiative, however, ended with the indefinite strike and State of Emergency—shifting from social warfare to military occupation of the Valley. Southern "pulled everyone out," recounts Aranda, "we did not do anything in the Valley, but what we decided to do was a very strong information campaign outside Arequipa, we did it in the rest of the country." Aranda did interviews with newspapers, radio and television stations, in addition to buying advertisement space in the rest of the country, until

> finally, the current governor said, "You should be doing this in Arequipa." The radios and the television said: "You should be doing this here." So then we came back, because one thing is for you to come back without anyone asking you and another is when someone asks you to return.

The main justification to return in 2015 came from the governor of Arequipa and the radio and television stations, the latter of which Southern is known to control. According to Romero, Southern retains "control over more than 13 local radios that exist in the Islay Province and Tambo Valley."[116] Reportedly, a brief case from a head journalist contained receipts for thousands of journalists that were paid by Southern on the quantity and quality of criticism lodged at mining opponents.[117] It seems that the media, and national and regional governments were the ones in favor of Southern returning—and much less so the residents of the Valley.

Then enters Valleunido in 2016, Southern's "community relations" third wave. Valleunido—Valley United—which consisted of 27 people responsible for the "information centers," going door-to-door every day with information brochures and implementing social development projects with the goal, in the words of Tía Maria's Social Relations Manager, to have "the Tía Maria project viewed as an opportunity and not as a threat." Largely recruiting people Indigenous to the Valley, Valleunido developed a corporate culture[118] that allows members to co-create the group name, brain storming sessions to better reach the population and a collective identity. "So we gave them information, we trained them," says Aranda, "we actually had training days for them where we explained what mining is, what the project was, how we do mining in our operations, what environmental concerns we have and what we do for those concerns."[119] This also includes pep rallies and measuring the impact of Valleunidos' efforts using census consultants, sociologists and anthropologists. Aranda, likewise, makes the distinction between "community relations" and "public relations." He explains that public relations "is mainly for interactions between our higher authorities in the company with the mayors, the church, social things like festivities," while community relations "are much more hands-on—working with the farmers, working with the cattle ranchers and the things that we do with the population."

Valleunido's community interventions fall under the later program: Construyamos Confianza Proyecto Tía Maria (Project Tía Maria Building Trust), which approached the community on six socioecological fronts. First, *Tambo Agricola* worked to improve soil quality, offering free fertilizers, pesticides, and "high-quality" seeds, in addition to classes on "improved" rice growing techniques. Additionally, Southern repaired irrigation canals and promoted mechanization of agriculture techniques. Second Tambo Ganadero was a program geared toward improving cattle and livestock by offering educational workshops, free straw, staffing veterinarians, and offering nitrogen tanks to allow the optimal conditions for cattle insemination. According to Southern, this project has "improved the quality of both the cows and pigs" with "an increment of 13 or a bit more increase in milk production."[120] Third, Mejora tu Vividenda offered portable water tanks and concrete floors for homes, while also renovating water infrastructure in select towns. Fourth, Apoyo a la Educacíon provided school materials, computers, uniforms, after school programs and painted and repaired parts of the school. Fifth, *Apoyo a la Salud* invested in medical clinics, 24-hour medical professionals, dentists, educational classes, and, even "has paid or is about to pay 25 million soles" (approximately USD 7.6 Million) for three studies for a hospital in Mollendo. Finally, community interventions also included three *Oficinas Informativas* (information centers) in Punta Bombón, Deán Valdivia, and Cocachacra,[121] which were located on main streets and, in Cocachacra, behind the

PNP station. These information centers were akin to internet cafes, except they had staff ready to discuss the benefits of the Tía Maria project, have you sit and watch Southern's promotional videos as well as handout brochures from Southern, the MEM, PNP, and Jehovah's Witnesses.

There have been three Valleunidos impact reports that, according to Aranda, can be summarized as: (1) "They do not like you"; (2) "We showed improvement ... [but] Yeah not so good"; and (3) "This last one is a bit better. We are actually ... pinpointing areas where we have to improve." Southern uses social scientists, anthropologists, and, more frequently, sociologists in an attempt to guide development interventions and measure their impact on convincing the population to accept the Tía Maria project. While studies revealed Deán Valdivia as being irreconcilable, these operations are consistent with research findings discussed in Chapter 2, but also how anthropologists and other social scientists measure the impact of the resettlement and development programs in Vietnam and elsewhere.[122] Speaking from the hydrocarbon zone in Ecuador, Wilson reminds us "The Oil Ministry is full of people ... with postgraduate degrees in Sociology and Anthropology, and training in the most effective means of sowing discord and engendering co-optation within anti-extractivism movements."[123] In this case, however, social scientists assisted Southern by measuring the impact of information campaigns to perfect a strategy to engineer the social acceptance of the Tía Maria project. "The engineering of consent," Edward Bernays explains: "learns what group leaders know and do not know, the extent to which they will cooperate with him, the media that reach them, appeals that may be valid, and the prejudices, the legends, or the facts by which they live."[124] This demonstrates how (often banal types of) knowledge can be weaponized and how social scientists are instrumental in executing social warfare and engineering the social acceptance of extraction. Success is measured in the lack of vocal opposition and protest, which as Aranda has observed is in decline.

A decisive political technology, according to Aranda, was the creation of "community committees," which decide what is important to their communities and what needs to be solved. Aranda, summarizing their initial formation, explains: "Listen, do not come and tell us that you need an airport that is out of the question, but if you see something that has to do with health, education, environment, culture, or whatever, bring it here."[125] Community committees allowed Southern to be strategic with their distribution of funds and create a participatory culture. "Central to controlling local communities," Hochmüller and Muller remind us, "is 'community input,' which can be translated into civil-military intelligence."[126] This rural development strategy deserves acknowledgment for addressing the Valley's social needs, but is unfortunately employed as a weapon of persuasion in the service of resource acquisition, extractive profiteering and "to have Tía Maria working."[127]

These counter-insurrectionary social interventions seeking the population's approval were intelligent. Nevertheless, the socio-political damage was done. The local response to social development was largely negative: "It's a scam," "Blackmail," and its purpose "is to change our mentality."[128] Another woman says, "the 'help' are like small pills to calm people down."[129] The portable water, concrete floors, fertilizer and so, required a signature that the people were convinced was being collected to show the MEM as proof of "social license" to operate. For example, one woman contends that Southern has

> a strategy, they say "I am going to give you a floor in your house, but you have to sign this piece of paper with your identification card" ... So they did not give you things because it is a gift, "no," they did it in order to collect signatures to bring it to Lima and present the documents that the people signed here to agree with the mine, but "everybody knows that this is blackmail taking advantage of people's needs."[130]

While a Valleunido canvasser denies this,[131] people felt that Southern was taking advantage of people's needs. This is reminiscent of Karl Polanyi's observations of "hunger," as instrumental to forcing people into work,[132] or Milford Bateman's (critique of microfinance as "poverty-pushed entrepreneurship"—where abject poverty pushes people into microfinance schemes because there is no other assistance or alternatives to development.[133])

While organizing deprivation is a classic technique of colonial counterinsurgency, so is the deployment of social amenities and gifts to get people to acquiesce to political control—convenience, luxury and stimulants, from technological devices to alcohol, are fierce weapons of social warfare. It was widely recorded, however, that Tambo Valley residents would take the benefits, but remained in complete opposition to the mine.

Human terrain studies, public relations and engaging in social development praxis, as it combines with state repression, also exploit local prejudices. The first was popularly conceived racism, dividing owning farmers from day labors. The racist discourse was especially salient in La Punta, among older *Criollo's* and municipal administrators. One municipal agent states:

> The farmers, the real native farmers, they are not the people who made the strikes. The people who made the strikes are the workers of the farmers, they are the people who work every day in the field—the people who fill their pockets with agricultural money.[134]

The day labors are typically migrants from Cusco, Puno, and elsewhere in the highlands,[135] which are home to large Indigenous populations. This view was tied up with a conservative welfare narrative that disparaged the communal food system Payapando and claimed that the day labors receive welfare and do not bear the agricultural overhead of tractors, fertilizers, and pesticides. This narrative creates a division between migrant day labors and owners, and, additionally, combines with claims that the Espartambos are agent provocateurs from outside the Valley. This external provocateur narrative is a textbook disinformation tactic across the world,[136] which according to many residents is completely untrue since "the Espartambos are from here, they are sons of Tambeños." Research participants repeatedly claimed that "everyone was united"[137] and "everyone from here went to the protest, it was all the people who worked in agriculture."[138] One teacher explains that "there are a lot of racist people," meanwhile another woman says Southern "take[s] advantage of this … they manage all of this,"[139] suggesting that they are exploiting a discourse that blames opposition and/or combative self-defense on darker skinned day labors and not the so-called "real native farmers." Counterinsurgency violence, social development, "outside agitator" narratives and racism, remain enduring doctrines and discourses seeking to divide, conquer and pacify people, meanwhile killing, degrading, and extracting from the ecosystems from which they live and subsist.

TÍA MARIA TODAY

The violent coercion and soft technologies of social war have been extensive, yet, as of this writing, the Tambo Valley has successfully prevented Southern from beginning operations—from 2009 to 2023. Recent turmoil in Peru, however, is generating concern. Since 2018, there have been many ebbs and flows. In December 2022, popular protests led by *campesino* and Indigenous movements have swept Peru after former President Pedro Castillo was impeached following his failed attempt to dissolve the legislature and his vice-president who, among other reasons, sought to overlook those allegations of corruption. The more conservative vice-president, Dina Boluarte, took over the government. On 14 December the Defense Minister, Alberto Otárola, declared a State of Emergency, suspending freedom of assembly, freedom of movement and other rights. This only aggravated protests and, on 18 January 2023, popular movements from the south of Peru marched on the capital in a mobilization known as the "Taking of Lima." Various demographics would join the protests, together demanding new elections for the presidency and legislature. This coup d'état and the subsequent uprisings have resulted in over 60 people being killed and thousands suffering injury.[140]

Known for his working-class background, Pedro Castillo was popularly referred to as the "anti-mining" president. After all, he grew up near Peru's largest gold mine in Cajamarca. The Tía Maria conflict was among the reasons that won him the popular vote, which offered hopes of a final termination of the project. Even in September 2021, when she was vice-president, Boluarte declared publicly on television: "Tía Maria will not go because it is unfeasible."[141] Since the political takeover, however, the

Boluarte government has already met with Southern twice. On 6 January 2023, the Minister of Energy and Mines (MEM), Oscar Vera Gargurevich, met with two representatives from Southern: Vice-President Raul Jacob Ruisanchez and former Minister of Defense (2020–2021), Nuria Esparch Fernandez, who now serves as the company's Senior Manager of Institutional Relations.[142] Then on 23 December, on one of the most important radio stations in Peru, Boluarte declared:

> I have been to Tía Maria just when we went to the presidential debate. I have been up in the Tambo. I have talked and met with several leaders and I believe that we have to talk with them … We have to talk, nothing is closed, but we have to respect life, and we have to respect the environment. Water is life. We cannot put the issue of gold, silver, above water … but also, I believe that the Tía Maria issue hurts the deaths (today), there are leaders there who have died in that struggle to defend their place, their environment. We need to talk.[143]

Then eight days later, Oscar Vera Gargurevich, from MEM, discussing the Tía Maria project told the same radio station that:

> We have to arrive with an understanding, explaining well the benefits and as we all know, we advocate progress and growth within a framework of sustainability in which the environment is mainly taken care of. Technically all of this is feasible, I think we have to reach the people who are complaining where we can demonstrate that we ourselves are the most interested in taking care of the environment and their productive activities … it is a matter of implementing our proposals and continuing to dialogue with all stakeholders.[144]

It appears the cycle of political promises and betrayal, identified by Tambo residents, continues, except this time with an empathetic appeal to "talk" and "continuing to dialogue with all stakeholders."

Other voices feel it is a waste that Arequipa only has one copper mine.[145] Some, like the governor of Arequipa, contend: "If there is not internationally audited environmental study, there is no point in talking. Because we know how environmental studies are done in Peru."[146] Representatives in the Islay province are firm in rejecting Boluarte having control over the process to begin a dialogue to start the process of opening the Tía Maria mine. People, however, avoid taking defined positions in general because of the politically tumultuous situation, yet it appears the Boluarte administration is preparing to advance a politically repressive and extractivist agenda in Peru. In short, while the Tambo Valley has endured state terrorism, entrenched hardship and political victory, the struggle to protect the Valley continues today.

Meanwhile, back in Arequipa in early 2018. I was writing profusely in a hostel. I had some friendly warnings from Southern Copper to "be objective" and "stop proselytizing" people, which was a nice way to say: "We are watching you." Given the intense interviews and insights, I was working on putting together a draft article, before leaving the country in order to be prepared if anything went wrong—computers disappearing or what not. During this obsessive, and highly regimented, writing period, I was contacted by people living in southern France. Land defenders in southwest France had taken an interest in my work in Oaxaca and Germany, and explained to me how they were fighting a mega-energy transformer in the Aveyron region. I was invited to present my work to people in the region, which unexpectedly initiated a long-term relationship investigating energy infrastructure in the region and beyond, which is the subject of the next chapter.

5

Trapped in the grid in southern France and Iberia: energy infrastructure and the fight against green capitalism

Following up on the invitation I received in Peru to visit L'Amassada in southwest France, it would only be a couple of months later that I arrived at the Montpellier train station. The renowned historian of science, Christophe Bonneuil, kindly picked me up and we drove an hour and a half north.[1] We entered the Aveyron region, going off the highway before the Millau bridge and then drove past the town of Saint Affrique and into the winding little mountains. Once we reached the top of the plateau, we drove along a ridge surrounded by large rolling fields, trees, animals grazing, and, to a lesser degree, wind turbines. Driving for another ten minutes or so, we arrived at L'Amassada, the protest site built in 2015 to oppose a large-scale energy transformer promoted by the French state and the Electricity Transmission Network (RTE).[2] At this moment, in March 2018, they were just finalizing a new wood-framed dormitory with recycled windows insulated with straw bales and mud, known as cob construction.[3] This was the second structure following the first created in the winter of 2015, which also had a small-scale wind turbine to power the site. Once arriving, I was able to meet the people I had started corresponding with in Peru and find out the details of my talk that evening, which I found out was going to be in the Saint-Victor-et-Melvieu town hall, about a three-minute drive from there.

That evening, the town hall was more or less filled with a diverse composition of people. I gave a standard presentation discussing my findings from Oaxaca and Germany and the similarities between them. Besides the diverse age demographic in the audience, what I found charming were the bright and exuberant faces who in the question-and-answer session, more or less, started making fun of me for how standard the presentation was—they wanted deeper and combative insights. I was not prepared to discuss such matters in a town hall, let alone with people I did not know. At the time, I did not expect this level of combative passion or anger from people who looked to be from an earlier generation and could be classified as "old" or, better, "wise," but were clearly as young and fiery as ever. Yet, when referring to the socioecological impacts, strategies, and tactics of the government, a woman blurted out: "The same thing is happening here!" In the following days, I was able to learn about the L'Amassada struggle, which means "assembly" in Occitanie, the traditional language of the region. My attraction to the project was that L'Amassada was at the forefront of the struggle against green capitalism, not only immediately seeking to prevent farmland exportation for a large-scale energy transformer—that would trigger more wind and solar projects to get built in the area—but also because the transformer was a key node in an energy highway, extending the grid capacity that would improve the transfer of energy across 400,000-volt high-tension power lines. This energy transfer is instrumental in further developing the European energy market, which— surprisingly—extends to Israel, Iraq, and the Western Sahara.[4]

The invitation to speak, the inter-generational rebel spirit as well as the intelligence of people present, not only in identifying but also taking action against the capitalist and extractive realities of low-carbon energy development, immediately sparked my interest. The following week, I would begin a process of interviewing people which would, eventually, lead to a three-year collaboration with the movement to further investigate the numerous issues L'Amassada was protesting and fighting. This transformer, and the movement against it, was insightful, demonstrating how a small town in the wet and cold hills of the Aveyron had regional and international consequences. This movement, and others along the 400kv power line, were generating action and public discussion on the reality of energy transition, so-called renewable energy development and the socioecological consequences of European environmental policy—specifically enthusiasm around the Green New Deal or

European Green Deal. The struggle against the mega-transformer, it turns out, was linked to other, and some older, struggles to fight this same energy highway. This also includes struggles against new wind (and solar) projects that would be hooked into the grid to promote the growth of the European energy market.

Figure 5.1 Barricades on a path to L'Amassada, March 2019. Source: Author.

WELCOME TO THE AVEYRON: A BRIEF HISTORY OF TERRITORIAL STRUGGLE

Aveyron is a county within the Occitanie region of southern France. This rural area, historically agrarian, "suspicious" of foreigners and regarded as "backward" by government bureaucrats,[5] consists geographically of small rocky green mountains interspersed with lively streams and the Tarn River. Known for inhospitable environmental conditions, this region relied exclusively on an agrarian subsistence economy.[6] The agrarian culture continues today, although economic activities have diversified into tourism, slaughterhouses, artisanal crafts, and service industries. The region is

France's largest producer of sheep, where dairy products and beef account for 40% of the agricultural market.[7] The Aveyron is home to the world-renowned Roquefort cheese, which remains an essential source of employment in the region. Energy production has relied primarily on hydroelectric resources and since 1999 has become a desirable location for wind energy development. Since then, as Alain Nadaï and Olivier Labussiere tell us, "there has been a growing opposition to wind power," an opposition that has only intensified over time.[8] Determined political struggles, however, have a much longer history in the region.

On 28 October 1971, French Defense Minister Michel Debré announced the 14,000-hectare expansion of the Larzac military base in southern Aveyron.[9] The base expansion threatened the expropriation of sheep grazing land—many associated with Roquefort industries—and triggered what would become a ten-year transnational campaign. Anti-militarism intersected with local territorial concerns on the Larzac Plateau. People migrated from all over France to support the struggle and it became a site of international solidarity, serving as a precursor to transnational "summit hopping" associated with the anti-globalization movement.[10] Inheriting and advancing Larzac's legacy is the ZAD (Zone-to-Defend) movement, but also numerous other territorial struggles in France.[11] There are, however, two important influential aspects of the Larzac struggle. The first is the migratory and circulating transnational solidarity networks of resistance, among them anti-mining and airport struggles in New Caledonia and Japan.[12] The influx of people coming from outside of the region was crucial, as "[l]ocal farmers exposed to outside campaigners became more cosmopolitan, open-minded and ready to contribute to wider struggles."[13] An important observation, Gildea and Tompkins note, was that the "success of the Larzac campaign depended on the sheep farmers co-operating with outsiders who had more organizational power and experience."[14]

The Larzac movement became legendary. Producing a generation of activists, notable among them José Bové who dismantled a McDonalds in Millau and eventually became a European Parliamentarian; this act served as a foundational inspiration for the anti-globalization movement.[15] The anti- or alter-globalization movements confronted corporate exploitation and international financial institutions with campaigns and shutting down international summits.[16] The limitations of the anti-globalization "summit hopping" model were felt,[17] leading toward locally oriented resistance of which by the late 2000s the ZAD movement would slowly become an exemplar.

In French, the term ZAD originally stands for "deferred development area" (*zone d'aménagement différé*). This signified the demarcation and reservation of land for a development project. The first recognized ZAD was in Notre-Dame-des-Landes. In the mid-2000s, a 1970s airport plan began to be implemented, resulting in various segments of the population organizing to thwart the displacement and defend the farmlands of local landowners. People refused to leave and began converting the so-called "deferred development area" into a "Zone-to-Defend" (*zone à defendre*).[18] Refusing displacement and land transformation, people not only sought to resist the airport construction by living in permanent ecological conflict, ready to defend the area against police invasion and eviction, but also initiated a prefigurative autonomous project of collectivity, ecological protection, and food autonomy.[19] It was not just about the airport, but as the slogan goes: "Against an Airport & its World," a world predicated on capitalist relationships and destructive socioecological practices. Permanent ecological conflict extends past ecosystems themselves, confronting the ideology and operations of land enclosure, extractivism, and capitalism itself, which views the world as a resource to be plundered in the name of "progress," "development" and "modernity." The Notre-Dame-des-Landes ZAD lived through multiple police-military occupations, and demolitions and succeeded in terminating the airport project in 2018.[20] The ZAD concept simultaneously spread, igniting a movement of permanent ecological conflict through autonomous land struggles blocking development projects, that *Le Monde* newspaper, in 2015, claimed totaled 27 in France.[21] ZADs were emerging to fight high-tension power lines, highways, dams (Sivens), nuclear waste dumps (Bure), ecotourism (Roybon), and more.[22] While resistance movements like the ones in Larzac and the anti-nuclear struggle in Plogoff (1980s)[23] remain important inspirations, the Zapatista struggle in Chiapas was also a foundational influence.[24] From the Narritia anti-airport struggle in Japan[25] (the 1960s) to Álvaro Obregón/Gui'Xhi"s struggle against wind parks in Oaxaca discussed in Chapter 2.

The ZAD concept has a profound affinity with generational territorial struggles across the world, of which the L'Amassada anti-transformer struggle falls within this constellation of autonomous land defense (see Figures 5.1 and 5.2). What makes L'Amassada unique, however, is that it challenges green capitalism, or "green" extractivism, as opposed to conventional infrastructure and extractivism projects. European environmental policy, and later the European Green Deal (EGD), claim that promoting low-carbon infrastructures, new digitalization schemes (e.g., "smart" censors and applications), battery storage technologies, and interconnections with high-voltage powerlines (HVPLs) to transfer solar, wind, and hydrological energy, will usher in energy transition and mitigate climate change. The policy claims have serious implications and, one might say, are gambling with the lives of people and ecosystems. The people in L'Amassada and in other HVPL conflicts, see this policy as misleading and to the det-

Figure 5.2 L' Amassada April, 2019. Source: Author.

riment of social ecologies, which has resulted in active resistance against them, raising the "red flag" and challenging the rolling out of the green economy.

BUILDING A NODE ON AN ENERGY HIGHWAY: AVEYRON, FRANCE

The transformer conflict began when the mayor of Saint-Victor-et-Melvieu, in 2010, violated town council protocols to approve the project. He also failed to adequately notify residents of the plan and its implied expropriation of farmland from specific families. A civil society group (Plateau Survolté) was formed, initiating legal battles that eventually led to (re)claiming farmland anticipated for expropriation by the French state and the Electricity Transmission Network (RTE). "I have studied various farming systems from permaculture to large-exploitative systems and from this perspective I want to decide how I want to work the land,"[26] explains Victor, whose farmland is being expropriated. Marie-Bénédicte Vernhet, Victor's mother, presenting her concern, explains:

> When you see the advertisements on TV saying: "our energy is green!" Is it "green?" We do not experience it as green because everywhere we see massive amounts of concrete coming in for the wind turbines and this awful transformer that is being built on top of the power lines and high-voltage lines we already have. We already had cancer in the village around here. So what will it be like with a big transformer? We feel like complete victims as well because the [old] mayors and all of them have decided first that they will allow these projects to settle here without asking us, let alone telling us.[27]

People were opposed to this project for numerous processual reasons, but also because, as Jean-Baptiste Vidolu contends: "It is not going to replace nuclear power, it is going to be added to it."[28] There are reoccurring issues related to political procedure, health issues, land-use and ecological factors—all of which motivated resistance to the transformer and nearby wind energy projects.

Among the procedural issues mentioned was the issue of propaganda and misinformation by the regional government and RTE. Police invasion and expropriation were looming, meanwhile energy transition and energy development were being promoted by the Regional Park. "In St. Affrique they do not talk about this transformer and it is the center of all these projects," explains "Jennifer," who had been attending a series of consultations on southern Aveyron's "Climate Plan" hosted by the Regional Park.[29] Attendants associated with L'Amassada accuse representatives of total denial regarding the impact of the low-carbon transition, which entails neglecting resource extraction supply webs, proposing inadequate mitigation measures, and intensifying socioecological problems by "greenwashing" regional industry. Concerned citizens and L'Amassada participants have been attending meetings and consultations since 2010. Between 2015 and 2018, L'Amassada organized to disrupt and shut down RTE and Regional Park events. De-escalating their interventions in 2018, their consultation participation could be described as a more "respectful approach." The structural concerns raised by locals regarding the wider socioecological impacts related to this development pathway, have been systematically excluded and ignored.

Protesters used to shut down meetings with physical disruption and blockade techniques, whereas now there is an emphasis on dialogue. The gymnastics performed by the Regional Park and political representatives to avoid and deflect criticism was shameful. The consultation meetings are usually advertised through newspapers and Jennifer remembers going to the St. Affrique City Hall "and two Regional Park engineers were welcoming us with games, similar to Monopoly, as a way to explain to us how great the ecological transition is." Many participants were insulted by this and perceived it to be "a way for deputies to publicize their [environmental] commitments" without addressing green capitalism and industrial expansion.[30] The relatively patient crowd listened to Chavillan who presented the board game "about land management" (see Figure 5.3). Haphazardly theatrical, this game promotes a reductive techno-managerial perspective by gridding the territory, organizing it for energy extraction and transmission, while situating the player in the role of energy developer. Thus,

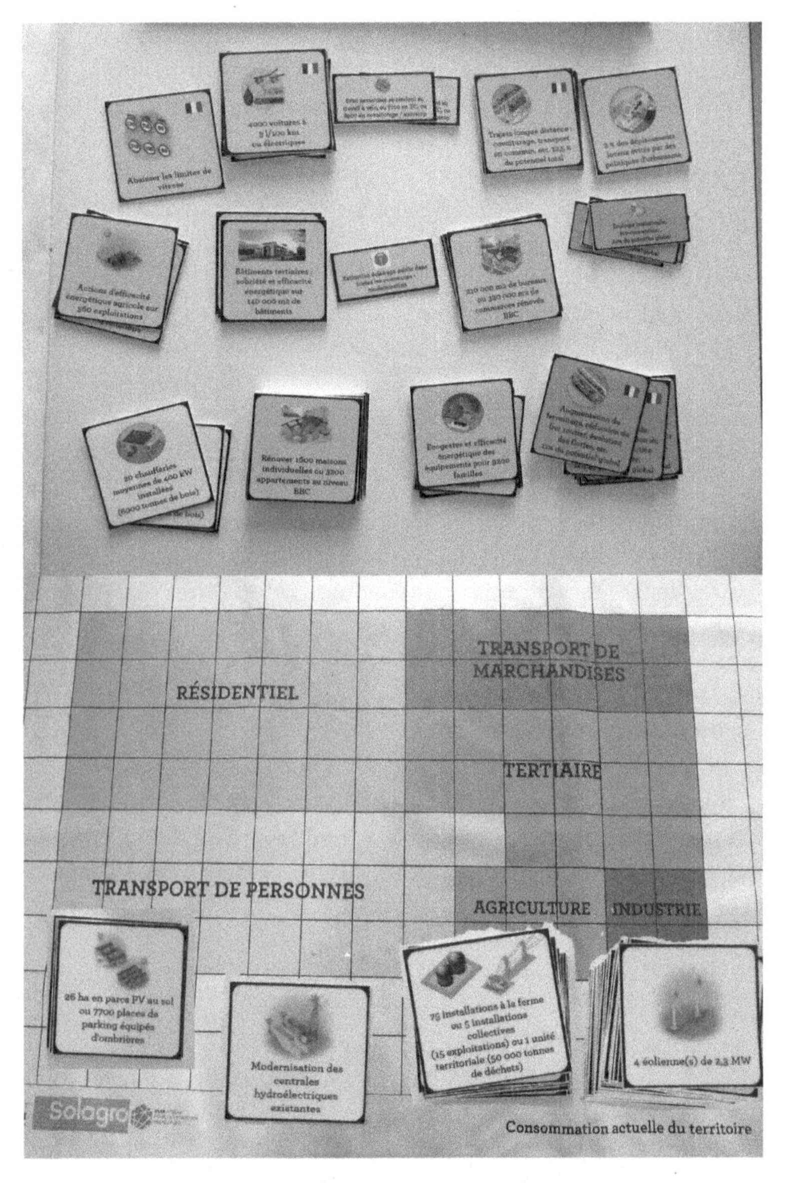

Figure 5.3 The board game of the Parc naturel régional des Grands Causses. Source: Université Rurale.

creating an opportunity for people to see like a state and regional planner in matters of energy development. People protested: "Are you kidding!? We do not want to play games, we want to see officials." Another person exclaims: "The absence of representatives is a continuation of the first meetings" that have failed to create agreement or resolution.

The Park representative, Chavillan, reminds the crowd that the public consultation "decree does not indicate how this consultation is supposed to take place." Said differently, people get a consultation but they do not decide its form or content. This further antagonizes the crowd, meanwhile Chavillan, supported by a parliamentarian attending the meeting, pleads for attendees to watch the proposed video the Regional Park has been trying to present. Eventually, people comply and humor the representatives. The video then proceeds to frame the issue of global warming, repeats existing statistics about energy consumption in the region—"58% of our consumption" was covered by renewable energy in 2017—and affirms the Regional Park's commitment "to cover all of our energy needs thanks to a renewable energy production by 2030."[31] People were outraged by the video, and how representatives are avoiding the responsibility of providing complete information regarding the impacts, supply chains, and economics of low-carbon energy development, while dispensing faulty statistics that included the A75 freeway (crossing between main metropolitan centers) that locals claim creates a negative statistical slant to justify the transformer and the spread of so-called destructive sustainable development projects. Furthermore, residents confronted Chavillan by identifying lies concerning the wind turbine numerical cap in the Regional Park—a cap that was placed at 300 wind turbines but has been exceeded. People felt betrayed by the process of decision-making without public consultation that had taken place and sensed that every energy development plan "is thought and decided before people even know about the projects." Chavillan keeps encouraging them to play the board game, justifies the statistical arrangement, and reminds everyone "that [the] representatives you criticize were democratically elected."[32] The consultation ends in theatrical and pointed fashion:

> Citizen (A75): We came here to tell you that we've had enough of your games and tricks and that we are always losing something because of your projects.
>
> [AC seems outraged]
>
> Ex-official: You came to this meeting to drag officials in the mud. You've done this for previous public events related to the establishment of energy infrastructures. You are discrediting your own movement/claims.
>
> Citizen (A75): The problem is that democracy doesn't exist.
>
> Ex-official: And it's because of you.
>
> Citizen (A75): I've been fighting industrial wind energy plans for 14 years now. We've managed to prevent the biggest project from being realized in Aveyron. But now we are talking about a wider scale plan, encompassing not only Aveyron but the entire Occitanie region. The fight goes on, but representatives refused to be with us tonight, instead of that they organized a [private] meeting on their own at 5 pm.
>
> Woman: We do not need your project, we are already self-sufficient regarding our energy production. You are developing renewable energies in an uncontrolled way.
>
> [People start to leave]
>
> AC: "You've ruined the meeting, thanks a lot."
>
> Citizen: No we haven't, we have raised a major issue regarding your plans and we must talk about it during a real public consultation.

The desire for adequate scientific assurances regarding energy consumption, local pollution, energy development regulations, the supply chain of raw materials and proposed energy use are denied by theatrical subterfuge.

Currently, so-called renewable energy supply chains are gaining greater attention,[33] and their dependence on hydrocarbons and extreme mineral extraction have earned lower-carbon infrastructures the title of "fossil fuel+."[34] In the EU alone, and based on 2018 demand, solar and wind technologies will (under a "high demand" scenario) increase the EU's demand for lithium, dysprosium, cobalt, neodymium, and nickel by up to 600% in 2030 and up to 1500% in 2050. Moreover, according to the International Energy Agency (IEA, 2021), offshore and onshore wind turbines and solar have a significantly larger material demand than nuclear and thermal power.[35] That being true, we cannot ignore the qualitative dimensions of uranium and coal extractivism, which present catastrophic ecocidal realities and concerns at every stage of their lifecycle. Furthermore, it cannot be forgotten that fossil fuels and nuclear power generation are inseparable from lower-carbon energy development in terms of stabilizing the energy grid and providing the energy necessary to mine, manufacture, transport and decommission fossil fuel+ technologies. The term fossil fuel+, again, seeks to highlight this industrial supply-web interconnection and integration, which is frequently framed by governments and mainstream environmentalists as opposites.

Fossil fuel+ systems have also become accomplices of natural resource extraction, not only in rebranding company images, but also in powering mining operations themselves, this includes the Agnew gold mine in Australia, the Aitik copper mining area in Sweden and Kevitsa nickel mines in Finland. The list is extensive, with efforts to electrify and digitalize mining, (which has an entire magazine dedicated to this topic, *Energy and Mines*). The issues raised by a range of oppositional residents highlight serious socio-ecological impacts and developmental trajectories with geopolitical implications—securing energy and raw materials to propel green capitalist development. The consultation, like others taking place elsewhere, in Latin America for instance,[36] worked to disregard structural political and scientific issues concerning energy infrastructure. The board game, in the end, deflected these concerns, meanwhile serving as a tool to normalize the infrastructural development or dominate planning perspectives.

Health and transnational energy flows are significant issues for oppositional residents. Already living with nearby energy infrastructure, St. Victor residents had existing health concerns: "I'm mostly worried about health issues regarding the population," explains Mayor Capel, acknowledging later that electrical infrastructure "undoubtedly creates health issues, and we have several cases of tumors and Alzheimer's disease within the population."[37] While the argument that electromagnetic currents are causing health issues is contested, it also resonates with numerous studies.[38] The correlation is difficult to prove scientifically, though the health concerns raised by people living close to electrical infrastructures were similar to those in La Ventosa, Oaxaca, discussed in Chapter 2. Marie contends that these health issues combined with negative emotional experiences related to the transformer's development process and police repression against individuals are negatively impacting the health of people in the town. R.H. Bakker and colleagues observe that in research participants "that benefited economically from wind turbines, the proportion of people that were rather or very annoyed was significantly lower, as if wind turbine sound was differently valued by them compared to non-benefiting respondents."[39] This relates to an emotional and political aspect, if the turbines are thrust on individuals who are deprived of the choice and do not offer noticeable economic benefit, then this will generate greater aggravating effects among local residents. These effects, however, are not just psychological or somatic. I interviewed three retired EDF and RTE employees,[40] who worked with the companies for over 30 years and opposed the transformer project. They acknowledged the lack of social-collective benefits, limited employment, the "evacuation of electricity made from wind energy" in the region[41] and the overall negative health impacts related to oversaturating the village with energy infrastructure. The interviews disclosed personal health issues caused by physical proximity to electromagnetic fields.[42] Based on physics calculations, the European Council's precautionary principle requires people "to be at least one meter away per kilowatt installed," explains "Patrick."[43] For a "400,000-volt infrastructure, populations must be at least 400 meters away, which creates an uninhabitable area," explains Patrick, although "there are no laws or legal rules regarding the distance between electrical infrastructures and populations. It allows the state to do as it wants."[44]

Finally, regarding the construction of a transnational energy corridor, Patrick asserts that the transformer is "supposed to raise the tension to 400,000 volts." The ENTSO-E (2019) map already confirms the transnational importance of the St. Victor transformer, which connects to the Asco nuclear plant, but also hydrological, wind, and solar resources in the southern Catalonian Terra Alta region.[45] Documenting Terra Alta resistance against energy infrastructural colonization of the region from the 1950s to the present, Franquesa shows how energy transition in Iberia was a process of energy transaction and accumulation, not socioecological transition.[46] Energy-capital accumulation and "successive additions of new sources of primary energy" were precisely the concerns of Plateau Survolté and L'Amassada, which administrators and politicians deny.[47] In consultations and interviews, political representatives claim "fossil-fueled energies or nuclear is meant to disappear in the future,"[48] yet according to Our World in Data energy use is expanding, where low-carbon energy development is serving as an addition to other thermal and nuclear energy sources.[49] Harnessing anti-nuclear sentiments is central to developing wind and energy infrastructure, yet EDF is currently building a new—and 3.6 billion USD over-budget—nuclear power station in west England[50] and, at the behest of France, is preparing to build six new third-generation nuclear power reactors over the next 15 years.[51] Meanwhile, political representatives emphasize "local" energy production and regional "renewable energy solidarity," but the "Occitanie local" borders with Catalonia and talk about a transnational energy super grid between the EU and North Africa that has been in the works since at least 2006. Yet while Plan Desertec has been scrapped for the most part, the ethos of the plan continues in a slow and progressive manner through the implementation of international climate change benchmarks and the promotion of a "green energy transition," which we will revisit below.

L'AMASSADA FALLS, A JOURNEY DOWN THE ENERGY HIGHWAY

The resources and political repression dispensed against L'Amassada have been significant. The transformer struggle, 'Meredith' contends, is "a war of attrition" organized with the "logic to get people tired and disinterested in the cause."[52] This grassroots movement has focused on blocking construction sites and roads for wind energy development and organizing information awareness campaigns, demonstrations, carnivals, and conferences, while disrupting public events. "Pirate Miau" explains: "We are building community together—I wouldn't use the term 'occupation' or 'ZAD,' as these labels are those used by the media and the state to categorize us in a pejorative way."[53] Wind turbine saboteurs, operating in the region, criticize L'Amassada's strategy, advocating "a backlash based on our own desires against those in power" on the basis of "[m]obility, stealth and unpredictability to stop the operation of capitalist infrastructures."[54] This communique was attached to the burning of three wind turbines in the area, which also criticized the social orientation of L'Amassada that organized public and mass movement events, but, in the minds of the saboteur(s), failed to take meaningful action by incurring economic damage by rendering wind turbines and energy infrastructure inoperable. In short, the saboteur(s) felt there was a lot of talk and posturing, but not enough results emerging from the direct actions taken.

While there might be truth to this, it does not change the fact the French state has attempted to criminalize the ZAD movement, trying to position them as "extremists" or terrorists.

ZADs, among them L'Amassada, in reality are rather politically moderate, exemplifying legitimate political concerns of great urgency. Politically moderate in the sense that Zadists are not assassinating engineers, politicians, nor police, nor engaging in public bombings of malls, movie theaters, airlines, and other terroristic actions. Instead, Zad participants tend toward identifying capitalism and industrialism as the cause of ecological catastrophe, meanwhile employing a participatory, nonviolent, and self-defense-oriented approach. All the violence responds to impositions and attacks from project developers and the state, resulting in acts of vandalism, arson and people defending themselves against invading riot police. This slander against the Zadists is reminiscent of what Indigenous land defenders in Oaxaca experienced (Chapter 2), being called the "Indigenous Taliban,"[55] while "Zadists" have been named a "green Jihad."[56] This trend has only intensified in the United States. Peaceful festival goers and tree sitters supporting the fight to defend the Weelaunee Forest, in Atlanta Georgia, have been arrested by the police, while 23 of them were charged with "domestic terrorism." The claims had no legal basis and were presumably an act of indiscriminate retaliation for the burning and destruction of police infrastructure usurping the forest on 5 March 2023.[57] The struggle in Aveyron, like Mexico, Germany, and Peru, experienced strategies of police surveillance, harassment, legal charges and eviction.

Between 2017 and 2019, L'Amassada was subject to frequent helicopter visits. Marie explains:

> Once they flew above my house six times—six times, then L'Amassada, the little village, the little hamlet there and another house in St Victor [which are all people concerned with the struggle]—six times. And who pays for the kerosene, who pays for the pilot and what for? What are they checking?

Carbon accounting, discussed in Chapter 1, rarely acknowledges the operations of political repression. Fuel-guzzling helicopters are not ecological, and are primarily used for creating an atmosphere of fear. On a Saturday after drinking a beer, "Nemesh" recounts turning around to a helicopter: "And suddenly we hear a sound and just above the hill, just next to us, there is: 'rwwwwwwwhhh,' an elevating helicopter—just in front of us. And they are looking at us and they have cameras, that's it." The helicopter was roughly "20 meters" away to the point where "you see the faces of the pilots inside."[58] Frequent helicopter visits were accompanied by the presence of Gendarmerie Mobile Platoon Intervention Units (*Les pelotons d'intervention de la gendarmerie mobile*)[59] and police surveillance. Frequent highway stops were made to invoke collective punishment for resistance to the transformer. Likewise, there were "two cop cars with a truck and it had a giant antenna with a guy with binoculars," circulating around the area.[60]

Surveillance was matched with arrests and interrogations. Responding to civil disobedience actions against the Crassous Wind Park (approximately 6 kilometers from L'Amassada), police raided 15 houses in January 2018. The aggressive and humiliating action of the raid varied among houses, yet the apprehended suspects were all brought to different police stations between 1.5 and 2.5 hours away from their homes. People believe this was a strategy not only to prevent counter-demonstrations outside the police station, but also to create transportation issues for arrestees. Some people were arrested in front of the schools where they were dropping off their children, while others were arrested while in bed. A woman recounts:

> It was like seven o'clock or something, but then I heard a voice that said: "Open the door!" I was like, "Ahhhh … fuck off! I'm sleeping!" Then the man with me was like, "Answer it, it's the police." I was like, "Nahhhh." Then they opened the door, I was naked and I was like: "What are you doing here!?" The chief of the group handed me a letter and I said, "Just a letter? You come into my home at seven o'clock to just give me this fucking letter?" And they said, "No, you are coming with us."[61]

A woman police officer watched her get dressed, and she had to explain to her son why she was being handcuffed. Frightening her son, the police raid was an exercise of intrusion, humiliation, and capture. The intelligence service, however, took a special interest in arrestees with generational roots in the region. "The local intelligence agency of Aveyron's DGSI (General Directorate for Internal Security) came into the interrogation room," recounts Victor. He "tried to explain that he was the guy from the intelligence agency and that I did not have to be friends with other cops, but I could cooperate with him—leader with leader." Victor refused, and the DGSI agent continued to pressure him, saying:

> All your fellow fighters have nothing to lose, but you do. Think of it, you have a baby. You are about to be a dad and you are running your own farm with your dad. You will be alone, your friends will have left and you will still be dealing with these things.[62]

The police did everything they could to divide oppositional residents from each other and from neighboring residents. Resources were mobilized to repress opposition to energy infrastructure. This "war of attrition," Meredith explains "is a repression spread out over time, it gets exhausting and it demobilizes people. The police know this and they play with it, hoping that the movement runs out" of energy.

The conflict lasted ten years, with the expulsion of the L'Amassada ZAD (Zone-to-Defend) in October 2019. Barricades were lit, people attempted to repel the police, but the two armored transportation tanks, 200 riot police, and demolition machinery pushed people out and pulled people off the roof of L'Amassada.[63] This led to immediate landscape colonization whereby the land was bulldozed, and enclosed with a fence and security guards to halt repeated protests for land reclamation and clashes with riot police. The result was the construction of an enormous 14-meter-deep hole, retaining 40 cm thick foundations. In total, the area is the size of a small airfield. The farmland, and the eco-anarcho-autonomist community on it, was transformed into an enormous hole held up by giant retaining walls at the valley's edge. The transformer, we must remember, was pitched as "green," "sustainable," and crucial to building a European energy transition, yet little is discussed regarding the amount of mineral extraction and energy necessary for manufacturing or transporting this mega-transformer. The justification to call this transformer "green" or "ecological" is relegated to carbon accounting practices, discussed in Chapter 1, that promote low-carbon infrastructures powering green capitalism and the growing energy-intensive consumption patterns of cities. This transformer development generated intercommunal and social conflicts, requiring sustained police repression and surveillance to promote the rapid and continued growth of wind and solar projects in the region.

Heading South to Bouriège and St-Sernin

This transformer was directly linked via a 400kv power line to other sites of contestation. Approximately 114 km south and 51 km to the west of the St. Victor transformer, in the Aude region, there was a wind energy conflict, which had been in process since 2004. Twenty-eight wind turbines already exist on an adjacent ridge, priming residents to reject the construction of another six wind turbines in Bouriège and St-Sernin. Similar to St. Victor's, the struggle began with deceptive public procedures, which led to locals organizing and filing lawsuits. Between 2004 and 2014 there was a sustained legal battle between the Occitanie Environmental Collective (OEC) and the companies involved.[64] The company in charge changed four times—from Tencia–Ecotecnia–Alstom to, finally, Valeco—obstructing accountability.[65] Residents again identified the legal struggle as "a war of attrition." Reflecting on the legal process, a long-time resident in their 70s, Agnès explains: "It is, in a sense, a war of attrition, because they [the company] force you to get some lawyers, to get really into it and it means that you have to invest [time and energy] in it too."[66] In this case, a decade of legal proceedings, lasting between 2004 and 2014.

Meanwhile, residents wanted to preserve the visual landscapes, and historical ruins, as well as the local wine and traveler tourism. People knew that more trees would be killed while heavy equipment would occupy local roads; existing roads would be widened and extended; ancient ruins destroyed; and, finally, landscapes would be visually dominated by enormous machines with a propensity to kill large numbers of birds.[67] "In the end," a resident explains, "they ended up forcing their way in and destroying a great part of the forest and a part of a historical site to build a new road that would go straight up to the top of the hill" where the wind turbines would be built.[68] The wind project clashed with the local economy, culture, and ecology. Social tensions were high between towns, creating tensions between St-Sernin and Bouriège. In 2014, two mayors signed a 40-year lease to build the "Bruyère Wind Farm." The lease signed, according to the then-current Mayor of Bouriège, had annual rents for Roquetaillade of €270,000 and Bouriège/St-Sernin of €55,000, yet the lease was signed when the mayors were no longer in power.[69] The proposed "Bruyère Wind Farm" was a source of revenue for the municipalities, even if residents criticized the mayors for their poor negotiations with the company. The general legal and political process led Agnès to exclaim: "I'm 70 years old, and until all of this happened, I still believed in democracy and justice. Now it is over."[70] The corporatist gymnastics of imposing infrastructure fermented disenchantment, cynicism and rebellion.

The villages closest to the wind project were resolutely against the project, meanwhile ambitious politicians were in favor and opinions were divided in the towns further away from the wind project. In 2014, oppositional landowners began blocking the roads in response, which eventually transformed into a ZAD in order "to make the struggle permanent."[71] Formed on private land, the St-Sernin ZAD created obstructions by every means necessary, which included building a checkpoint that utilized and enforced property lines to prevent heavy machinery from accessing the project site.[72] "In the 48 hours after the Zadists' arrival," Agnès explains, "the cops were informed about their presence—they were only three in the beginning—and we started seeing police helicopters flying over the hill and the ZAD." Police helicopters and surveillance; visits from the Protection and Intervention (SPI) squad; [73] attacks from company security; and blocking inhabitants in their villages were all recurring features in the St-Sernin struggle. In the end, Valeco obtained a permit to build a costly road, and equipment entered the land with security agencies. Valeco private security tried to subdue land defenders, which accompanied a police escort for wind turbines as well as organizing €1 million legal fines for people who were preventing the wind factory construction.[74]

The road struggle lasted three years in total.[75] Police repression, similar to the St. Victor transformer conflict, employed surveillance (helicopters and personnel); visits from political police; attacks from company security and police; and the prevention of inhabitants from entering or leaving St-Sernin. The Bruyère wind extraction zone was completed in 2017. Currently, the wind park does not have a direct connection to the 400kv power line. The region, however, is under constant pressure to receive more wind and solar projects, leading George, a local degrowth advocate, to anticipate that "it is highly probable that the power lines in the region will be doubled or upgraded in order to export the surplus toward other cities or to the Costa Brava [Catalonia]." This lesser-known ZAD struggle reveals the lesser-known conflicts over wind energy development.

No a les Línies de Molt Alta Tensió (MAT): Catalonia and Spain

Directly connecting with the St. Victor transformer is the anti-MAT (very high-tension lines) struggle in Catalonia, central and southern Spain. The anti-MAT struggle exists alongside many fights against factories, tourist developments and highways.[76] The first 400kv AC (alternating current) MAT line between Vic–Baixas was built in 1982, with plans for an HVDC (High Voltage Direct Current) emerging in the 1990s. The new MAT plan led to the formation of the anti-MAT movement in the Girona region.[77] "[T]he liberalization of the electricity market and the ambition of the European Union" to meet lower-carbon energy objectives made the French-Spain Electricity interconnection (INELFE)[78] with 50% participation from RTE and REE (Electricity Network of Spain), an EU "Project of Common Interest."[79] The project had two phases between the Vic–Bescanó transformer (1999–2011) and the Bescanó–Santa Llogaia (2012–2015), which connected to an underground line from Santa Llogaia–Baixas. The "No a la MAT plataforma" formed in 1999 and was at its peak in the late 2000s. The plataforma, at its peak, in Girona had 8,000 people involved,[80] and began placing ecological issues at the center of political struggle.[81]

The MAT line had extensive socioecological impacts. MAT lines, ecologist Marc Vilahur[82] reminds us, induce habitat loss and fragmentation by killing trees with new access roads and HVPL towers. Soil was compacted and there were changes in vegetation structure and animal populations.[83] Ecosystems were disrupted, creating openings for "invasive species" or, more accurately, new species responding to ecological disruptions.[84] Bird populations are also severely affected by power lines. Another ecologist, Josep Bort,[85] studying MAT lines in the Spanish Valencia region in collaboration with other conservation groups, contends that "900 to 1000 birds of prey are killed every year by electrocution." Josep claims that in Castellón, "it is about 300 birds a year." Yet "most of these dead birds are found by farmers on their land" and "there is no systematic organization or project to search for electrocuted birds in the Valencian community." Funded research projects, Josep estimates based on studies, "would reveal numbers that are three to four times higher, around 4–5000 electrocuted birds." Reviewing power line impact studies on biodiversity, Biasotto and Kindel documented "38 studies that reported bird mortality by electrocution and 12 of them identified this as the main cause of population decline."[86] The ecological impacts of HVPLs are far removed from the EGD and GND conversation, which exist alongside the cumulative effects of conventional extractive industries, lower-carbon infrastructures (wind, solar, hydro, etc.), highways, tourism and others.

MAT infrastructure is a potent symbol of the colonial domination of Spain, led by the Francoist-born company of REE,[87] over Catalonia. MAT infrastructure and the corresponding energy market, colonize and visually dominate the Catalonian countryside. "You can't be independent if you are not independent in terms of energy," explains Sergi, "It was the Mossos [police] who imposed the supply of energy from Red Eléctrica de España." Criticizing large sections of the Catalonian Independence movement for ignoring infrastructural colonization, Sergi testifies that MAT infrastructure could only exist by coercive force. Resistance to the project was strong, which—despite popular concerns—was violently repressed. This became a process of engaging in permanent ecological conflict between 2007 and 2015, which included countless demonstrations, information events, and an anti-MAT forest occupation outside Sant Hilari Sacalm (2009–2010; see Figure 5.4), and continued with protests on construction sites, repeated "lock-on" actions, and site occupations, to prevent construction. There were countless sabotage and incendiary actions during this period of time. "Jaime" explains, "More than 100 different acts were done, but people didn't really hear about them."[88] Publicizing communiques can create safety risks for land defenders and requires detailed internet security to post actions. This militant opposition, however, was violently suppressed: "There were a lot of controls, the police made local people's lives a misery," explains "Calico," "They would stop you several times a day and would hold you up for hours."[89] Speaking about the police controls, "Vetti" remembers that "from August 2013 until about a year later, some local villages started to look like a Palestinian checkpoint."[90] Occupation by police in duration and intensity was significant. High legal fees, jail time and drawn-out legal procedures, lasting up to six years for civil disobedience actions. Central features of permanent ecological conflict, sabotage, occupation and civil disobedience actions were shared across all the sites.

Figure 5.4 MAT line passes above Sant Hilari Sacalm, the point of struggle between 2009 and 2010. Source: Author.

Despite the 2015 completion of the MAT line, the same year the St. Victor transformer was originally intended to be completed, a third MAT branch in Girona and more anti-MAT struggles in the Valencia and Andalusia regions of Spain sprung up. In Girona, No a la MAT Selva is currently organizing, and proposing alternatives to the MAT line, demonstrating its obsolescence and the profit motives related to high-speed trains, tourist developments, and casinos. Struggles are similar in Castellón and Granada. Information sharing, public consultation, the buy-off of local representatives, divisive land contracting tactics, and, overall, socioecological fragmentation are key factors propelling anti-MAT resistance. "In general, people who own the affected properties are not happy about the lines," explains Joaquin.[91] Summarizing the situation in Andalusia, Joaquin continues:

> At first, landowners are told by Red Eléctrica that there is nothing they can do to stop the project. Then, they are told that if they accept the deal, they will get a lot more money than if they are legally expropriated by the state. The majority of people accept the deal. Out of 1000 people, only 3–5 people refused the deal and opposed the project.

The struggle is stronger around Castellón, and oppositional resistance within public institutions has only gained strength over time in Girona. The 400kv line passing through Girona, Castellón and Granada, continues south, connecting with other above-ground MAT lines, but also two—and soon three[92]—transnational submarine MAT lines connecting with Morocco.

Consistent with the EGD's interest in transnational interconnections, RTE is planning the Celtic Interconnector Project, estimated at a cost of M€930 to build 575 km of 400kv (HVDC) line between Cork in Ireland and Brittany in France.[93] Morocco is embarking on a similar path.[94] In addition to new line connections with Spain, Morocco is considering a UK-based Xlinks project. This is a 3 GW submarine cable linking Morocco to the UK that, according to Boulakhbar and colleagues, "will allow green electricity to be sent directly to the UK without using existing infrastructure in Spain and France."[95] This project is intended to generate 6% of the UK's energy demand, which relates to the rapid development of lower-carbon infrastructures, power line reinforcement, and expansion.[96] "Renewable energy," Joanna and colleagues explain, "has the advantage of casting a favourable light on Morocco due to the power of greenwashing discourses."[97] Solar and concentrated solar power (CSP) projects designed on centralized grid systems, with little interest in including locals in the project, are a recurring problem for local people.[98] The struggle continues as the 400kv line links to contestations against land grabbing for wind and solar infrastructures in the Western Sahara.[99] Allan and colleagues, making an effort to speak with people not identified as "activists," find that wind and solar development led by Siemens in the Western Sahara "fuels Moroccan colonialism and occupation."[100] Mahjoub, a fisherman, explains that the wind turbines "do not represent anything but" the wind and "your land being illegally exploited by the invaders with no benefits for the people."[101] Siemens, developing five wind projects in the region, Allan and colleagues note "ignores the existence of [the] Western Sahara in its PR materials."[102] Ignoring the Western Sahara affirms the process of colonial erasure.

The Moroccan regime is embedded in colonial authoritarianism,[103] making resistance to green grabbing complicated and deadly. "Because owners or users of this traditional land are either displaced as refugees—half the people have been displaced since 1979—or they are terrified to protest, demonstrations are not allowed," explains a member of Western Sahara Resource Watch (WSRW). Infrastructure in the Western Sahara is weaponized, using electrical "blackouts" strategically to stifle resistance, reinforce "othering" and immiserate the Saharawi people.[104] The Western Sahara region ranks among the lowest in political freedoms in the world.[105] A WSRW member explains:[106]

> So the opposition you see to these wind farms—they are massive—but they are always in terms of political declarations from people who are not sitting on the construction site. They will either be in an apartment in the occupied territories, 20 km away, issuing a press statement, maybe with a photo or a banner.

There are, however, active protests, "sit-ins," and roadblocks against the energy projects, but this results in violent repression: beatings, job losses, benefit cuts, travel bans, and death threats.[107] Sultana Khaya, president of the Saharawi League for Human Rights and Natural Resources, who led several protests against specific energy companies, was eventually detained and had her eyes removed by police. Secret prisons are frequently used to thwart anti-colonial resistance, which is entangled with opposition to low-carbon infrastructures, where torture, rape, and forced confessions are standard tactics to suppress Saharawi resistance.[108] Even under these politically inhospitable conditions, there have been mobilizations and suspected acts of sabotage against energy infrastructure in the Western Sahara.

The degrading, conflictive and cumulative cost of energy infrastructure is multi-dimensional, multi-sited and under-acknowledged by European Greed Deal advocates, and political "greens" in general. The situation is well summarized by Alexis, who fights the MAT lines near his organic farm outside Castellón.

> Renewable energy is not produced in Morocco and is carried all the way to Belgium, even if it is produced by wind turbines. That is capitalism. Renewable energy is instead the self-production and local consumption of electricity. You can sell me ecological organic fruits from Peru and I can eat it here, but then it is not ecological.[109]

While degrowthers, environmentalists, and others embracing climate change and Green New Deal style policies can view them as an opportunity, its relationship to low-carbon infrastructures and "renewability" is questionable, to say the least. Green growth remains a myth, which becomes even more intense when we consider what low-carbon infrastructures are really designed to do, which is to generate profit and, this often implicitly entails, encouraging increases in energy demand within a capitalist or "growth" oriented system.

THE GRID INVITES CONVENIENCE, BUT IT IS ALSO AN EXTRACTIVE MACHINE

New EU environmental legislations, such as the European Green Deal, as mentioned above, are promoting new interconnections with high-voltage powerlines (HVPLs), grid reinforcement by upgrading HVPLs, and transformers, as well as explaining lower-carbon technologies such as wind and solar generation, digitalization (e.g., "smart" censors and applications) and storage technologies. This environmental policy approach is generating the projects, dilemmas, and conflicts discussed above, yet the greatest issue with these projects is that they are not resulting in a real energy transition or creating a genuinely self-sustaining socio-ecological renewability. In fact, quite the opposite. Public funds are designed to create and stimulate an energy market, which includes its physical infrastructures. We might conceive of transformers and power lines, especially high-voltage ones, as the physical energy stock exchange, because all of this is about buying, selling, and profiting on energy production—green or conventional. The EGD claims that climate change mitigation objectives "will require €260 billion of additional annual investment."[110] Green energy transition public funding mechanisms are extensive. This includes, first, the Connecting Europe Facility that invests €8.7 billion into energy infrastructure networks.[111] Second, the Recovery and Resilience Facility where "green transition" is one of six pillars with its public–private fund of €672.5 billion.[112] Third, the Financing Energy Efficiency scheme with €18 billion allocated to "sustainable energy projects" of "strategic priorities" in the years 2014–2020.[113] Fourth, the European Energy Program for Recovery was designed "to promote energy transition" with a €3.98 billion budget between 2009 and 2018. Fifth, the 2014 State Aid program that in 2016, for example, allocated €6.4 billion to various projects,[114] many related to energy infrastructure. There are other private sector loan schemes (e.g., European Investment Bank and the European Fund for Strategic Investments), yet there is an enormous amount of public funds dedicated to creating energy markets and allowing profiteering from energy consumption.

There are three different ways to trade electricity: a derivative market, a spot market, and over-the-counter trading.[115] Interestingly, spot markets rely on electricity exchanges, for example, EPEX,[116] which comprise an assemblage of digital platforms and monitoring mechanisms, "to handle the generation uncertainty of intermittent power sources."[117] In 2011 and 2014, this included day-ahead and intraday markets where trading could be conducted between various intervals up to 15 min.[118] The political goal of any market-based solution, explains Ocker and Jaenisch, is "welfare maximization," entailing optimizing "liquidity, market power resilience, information efficiency, static allocation efficiency, and efficient usage of cross-zonal capacity." [119] There are, however, already instances of technical error (e.g., partial market "decoupling" in June 2019) and the European Commission (EC) investigating EPEX Spot for violating anti-trust laws, such as "taking advantage of its dominant position to hinder the activities of competitors on the market for electricity intraday trading" in six member states.[120] Fossil fuel+ technologies intermittence, combined with consumer choice as well as (national grid centralized) wind and solar parks decentralization extends neoliberal political and economic objectives.

Transmission System Operators (TSOs)—the institutions responsible for managing the electrical grid like RTE—and their public infrastructure, manage and dictate the size of an energy stock exchange. Energy stock exchanges necessitate fossil fuel+ power plants; power lines, transformer substations, "smart" monitoring, and market interface technologies; regulatory agencies; and the socio-technical system of TSO administrators. As mentioned above is the enormous extractive, smelting, manufacturing, transportation, and decommissioning cost that remains completely ignored by European Commission and think-tank reports. Representatives from the Electricity Transmission Network (RTE)—France's TSO—stress, "We just offer exchange capacity" between "suppliers and generators,"[121] where the prices of conventional and lower-carbon power are managed on electricity exchanges.[122] TSOs, like RTE and REE, are trading watchdogs ensuring energy exchanges are within "a 2 or 3% margin," "regulating prices and making sure the [energy] balance" between supply and demand is reached (avoiding grid collapse). "Greenness" and "renewability" are not the preoccupation of TSOs; instead, this is relegated to market logics via energy suppliers and certification schemes managed by TSOs.[123] "When someone buys green energy, we make sure that suppliers of this green energy have produced or bought it," explains an RTE representative, "it is a financial exchange, the buyer will pay for his green energy to the producer, and we just make sure it happens according to the market rules."[124] Central in this conversation is the necessity of the grid, as RTE representatives remind us: "You need the grid, because to reach these amounts of renewable production to compensate for a potential decrease of gas and nuclear, by chance we will have a sufficiently adequate network" by drawing on infrastructures nationally and internationally.[125] This references the imperative to strengthen national and international interconnections and spread of low-carbon infrastructure in order to create "more secure energy supply" and "better and safer energy markets." As we have seen, this EU approach to "green energy transition" has ignited conflict and decade-long struggles. Examining a segment of an HVPL between southern France and south-central Spain.

"THEY" ARE STILL PACKAGING OLD WINE IN NEW BOTTLES

The green economy as we learn in southern France and Iberia, but also in Mexico, Germany and, to a lesser degree, Peru, is just the latest adaptation of capitalism and extractivism. The "green" aspect we recognize is just marketing and the implementation, or greater emphasis, on new wind, solar, hydrological, and other extractive technologies. On closer investigation, it is the same energy infrastructure, extractive, and profiteering ideology that justifies itself through incomplete, partial, or misleading accounting practices, as discussed in Chapter 1. Power lines, transformers, and kinetic energy extraction technologies (e.g., solar, wind, hydrological) still require conventional extractivism. In the EU alone, and based on 2018 demand, solar and wind technologies will (under a "high demand" scenario) increase the EU's demand for lithium, dysprosium, cobalt, neodymium, and nickel by up to 600% in 2030 and up to 1500% in 2050.[126] Batteries for electric vehicles, wind, and solar technologies will drive the EU's demand for lithium up by 1800% and cobalt by 500% by 2030, and in 2050 demand will increase by almost 6000% for lithium and 1500% for cobalt.[127] These approximations, however, still do not take into account material demand for high-voltage powerlines, transformers, electric scooters, bikes, and other digitalization or "smart" censors, not to mention the increasing demand for electric vehicles in non-EU countries like Norway that have surpassed earlier predictions.[128] There is nothing socioecological, sustainable or renewable about the green economy as it currently stands. Only from the narrow, and limited, lens of decarbonization—from the perspective of carbon accounting discussed in Chapter 1—can any of these actions make sense. The perspective of decarbonization teaches us to think about the world in terms of carbon, relying on expert studies that are questionable, instead of responding to the real needs of our environments, relationships and ecosystems, which cannot and should not be reduced to just carbon. There is so much more going on with people and ecosystems.

I sought to extend my research to Morocco and the Western Sahara region, in the hopes of drawing connections. This, however, was cut short by intense COVID-19 restrictions between Valencia and Granada. My friends and contacts there warned me that there was only a 50% chance of crossing into that area and that I was likely going to have to put myself in the trunk of a car. This was unappealing, and the last three months of locating conflicts and conforming to COVID-19 restrictions along the 400kv power line, had already exhausted me. Yet, at this exact same time, I was contacted by people in northern Portugal. Having become informed about my work, people contacted me to tell me about their struggle against lithium mining. I was invited to the region, not only to learn more about the struggle, but to present and share what I had been learning and studying regarding energy transformers and power lines in France and Iberia. Eventually, I took a bus up there from Valencia and, like my experience entering the Aveyrone, I would continue to develop a relationship with the movements in the area and investigate lithium mining in northern Portugal, to which the next chapter turns.

6

When environmentalism is ecocide: an open-pit lithium mine, Portugal

We are in a small village in northwest Portugal. In December, it is surprisingly warm with spring-like weather. My friend and I are driving through small villages to get to the main highway on our way to Covas do Barroso to celebrate the New Year. The Barroso region is among the main areas targeted for an open-pit lithium mine, and among the largest proposed lithium mines alongside mines in Serbia, Finland, Spain, and France. Covas do Barroso is one of the towns on the forefront of resistance to the mine, which will be situated almost a stone's throw away on the edge of the village (e.g., Rominho). This New Year's Eve was not only to party, but to meet up with people organizing against the mine and converge with others in the country concerned about mining and environmental issues. We arrived at an old school used for communal activities, which the communal land authority allowed people to stay in for the week. Having previously testified at the European Parliament, at the "Environmental Citizenship, Public Participation and Transparency in the Mining Sector in the EU" hearing, with the help of Mining Watch Portugal I lambasted how the European Commission was employing public and Horizon research funds to create initiatives to educate and engineer local acceptance of mining.[1] People in the Barroso region were watching this hearing, which generated some interest in discussing their struggles and dilemmas with me. Not long after introductions, we made our way to a big bonfire and festivities. In a typical village style, the people would come together to be merry and forget the mining burden looming over them.

The following days were spent talking, hiking, and learning about how the communal water and irrigation canals were managed, visiting monuments and the communal house at the center of town. The area is famous for its livestock, agricultural practices, and its Indigenous breed of "barrosã cow." Any and all carnivores should be interested in the beef and pork in this area, which—unlike the supermarket—would be matched by the occasional and horrendous screams from pigs seeing their final days that would occasionally echo through the village on my visits. The Barroso region is recognized for being an important biodiversity hotspot in Portugal and home to endangered species. People here have preserved ancestral ways of working the land and rearing livestock, which in 2017 became Portugal's first area classified as a "World Agricultural Heritage" site and added to the "Globally Important Agricultural Heritage Systems" by the Food and Agriculture Organization of the United Nations.[2] This area is known for its communal agricultural system, which also operates on communal land, known as "baldios" or "terras baldias." This ecosystem and traditional agriculture, however, is being targeted for open-pit lithium mining or, said differently, *total degradation*. Large-scale and mechanized open-pit mines are the antithesis of traditional agrarian landscapes, even if the two tend to merge by force and people adapt.[3] The imperative for this lithium mine relates to European environmental policy, pushing to develop electric vehicles and energy storage systems that will store wind and solar energy. The belief in these industrial-scale and extractive-based systems being "green," "environmentally friendly," or "clean" will mitigate climate change and generate a transition away from fossil fuels to so-called renewable energy. This belief in industrial and digital renewability perpetuates the need to expand lithium and other mineral mines at unprecedented rates.[4] While, at the time of writing, the Barroso mine has not started operations, and, while it has lingered for some time in the advanced stage of permission, on 31 May 2022 the Portuguese government has given it the "green" light to become the largest lithium mine in Western Europe. This approval, however, has numerous "conditionalities" that will cause further delay.[5] While signs of permanent ecological conflict are fleeting in the area, the mine is not built and people are still organizing to defeat it.

Figure 6.1 Covas do Barroso. Source: Author.

BACKGROUND: THE RISE OF LITHIUM IN PORTUGAL

Since the Paris Agreement (2015), the European Commission (EC) has intensified its efforts to mitigate climate change and ecological catastrophe. Central to this strategy, as mentioned in earlier chapters, has been the rapid expansion of low-carbon infrastructures and the decarbonizing of the transportation section by developing electric vehicles (EVs). Lithium, cobalt, graphite, copper, and nickel remain instrumental to the lithium-ion batteries used in EVs and utility-scale energy storage systems, the latter of which enables grid stability by storing (intermittent) solar and wind energy. This socio-technical energy transition has placed lithium in high demand globally and within the EU. The EC wants to place 30 million EVs on the roads by 2030, ensuring that nearly all cars will be zero-emission by 2050. The EC currently maintains 100% import reliance for lithium in Chile (78%), the United States (8%), Russia (4%), and presumably China and Argentina (10%).[6] Seeking to break import dependency, the EC wants to produce 89% of the batteries for EVs in Europe by 2030,[7] and has thus initiated legislative and administrative efforts to intensify lithium extraction within its borders. Specifying local material content requirements—the amount of material mined originates from within home countries or the EU—are increasingly common within trade agreements. This entails measuring the final percentage of locally sourced materials with the final product, which requires a critical examination of how accounting is measured, calculated, and (epistemically) justified. Local requirements seek to boost national economies, but also—at least in words—redress the ever-present extractivist colonial relations and further justify expanding domestic mining operations alongside existing international mineral supply webs.

The EU currently allows 70% outsourcing of the total car value outside of Europe. In January 2024, this will be reduced to 50% of the car value, while batteries will need to retain 40% of local materials, according to Bridge and Faigen,[8] which "will effectively require automakers to source batteries from within the EU." As such, the EU now estimates the need for 18 times more lithium by 2030 and 60 times more by 2050 compared to 2020 supply levels.[9] It is no wonder Galp—Portugal's largest petroleum company—and Northvolt—Sweden's largest EV battery manufacturer—have now signed a joint agreement to open a "lithium conversion unit," or refining and manufacturing facility in Setúbal, Portugal.[10] Batteries for EVs and energy storage systems, the European Environmental Bureau (EEB) shows, "are predicted to drive up demand for lithium by almost 6000% by 2050" by 2018 standards.[11] Moreover, EV market shares are rising faster than expected in Norway.[12] This demand is transforming lithium into the new "white gold." In addition to the European Battery Alliance, established in 2017, the EC presented the Action Plan on Critical Raw Materials, implemented the European Raw Materials Alliance, and added lithium to the list of Critical Raw Materials in September 2020.[13] Recently, in September 2022, the EU announced the launch of a Critical Raw Material Act that permits the creation of common strategic projects of European interest, allowing these common projects to receive European funding and expedited approval processes. This policy and consumer trajectory places the Barroso region, or Portugal in general, in the mining crosshairs of the European Commission.

The Portuguese government views the current "green" energy transition as an opportunity to place the country in a position of leadership within the EU.[14] Lithium deposits in Portugal were first located in 1992 by two geologists from the French National Geological Services and Fernando Noronha, a Portuguese geology professor and researcher.[15] Since the early 1990s, the Portuguese National Geological Services institutions[16] have given priority to lithium inventory and research, with an increased intensity since the early 2000s.[17] Now Portugal is known for having Europe's largest estimated lithium (metal) reserves,[18] and is one of the world's top ten producers of lithium,[19] ranking seventh according to the United States Geological Survey.[20] In 2019, it had a 1.6% share of the global production.[21] Lithium production in Portugal, however, is used exclusively for ceramics and glassware, and not for electric vehicle batteries. There are currently no lithium mines in Portugal—rather, there are quartz and feldspar mining operations that collect some lithium—but the European green transition is expanding lithium development in the country, specifically for electric vehicles.

In 2016, the Secretariat of State for Energy created the "Working Group on Lithium." The group's aim is to make lithium *financially viable* by (1) identifying and characterizing the economic activities of exploitation associated with lithium; (2) establishing a hierarchy of priorities of industrial use to maximize economic benefit; (3) defining a program for the production of lithium compounds; and (4) proposing measures to substantiate the creation of a specific processing unit for these minerals.[22] Between 2016 and 2019, the government authorized requests for prospecting and research of minerals across 19.3% of its territory.[23] In 2016 alone, 30 applications for exploration and prospection rights for lithium, totaling an area of 2,500 km^2, were filed with the Directorate-General for Energy and Geology (DGEG), the main institution responsible for authorizing mining contracts.[24] The lithium "rush" has and continues to stir political and economic ambitions in Portugal. Data shows that almost 25% of the country's continental landmass might be reserved for mining projects.[25] European environmental policy ambitions, along with the objective of extractive enterprises, are spreading mining across Portugal and, likely, elsewhere.

THE EPICENTER OF THE LITHIUM RUSH IN PORTUGAL: THE BARROSO REGION

The northern region of Barroso is composed of two municipalities, represented by two City Halls (*Câmaras Municipais*): Montalegre and Boticas. This region is particularly rich in geological deposits and minerals, such as wolframite (e.g., a tungsten ore mineral), niobium, tantalum, and lithium.[26] In Montalegre, a significant wolframite mining exploitation existed in the small town of Borralha, from 1902 until 1986, which left concerns of environmental toxification (e.g., polluted water sources) and people jobless without remediation or compensation.[27] Today, it is possible to visit the Interpretive Center of the Borralha Mines, part of the Barroso Ecomuseum, which holds an important collection of documents about the mines. The studies conducted on the environmental impacts of the Borralha mines have concluded that all the areas (soil and water) are severely contaminated.[28] Despite—or partly because of—the visible socioecological degradation left behind, DGEG has signed a contract with Minerália in 2021 to reopen this mining site and exploit tungsten (wolframite), tin, molybdenum, and related metals in a total concession area of 382 ha. Governmental officials, and pro-mining actors, are mobilizing this mining heritage, by recalling the "economic boom" the mining operations brought to this historically vulnerable and poor region. This case demonstrates how territories depleted by extractivism can easily become (re)colonized for new extractivist purposes.[29] The cumulative consequences of mining result in ecological, social and economic *drain*, rendering these territories and their populations—human and nonhuman alike—more vulnerable to extractivist-infrastructural reappropriation.

The comparably high concentration of lithium reserves present in the region makes it particularly "promising" for production and exploitation.[30] In this region alone, eight mining exploration contracts have been signed, while seven mining applications are in the process of being analyzed by DGEG. In the last five years, 39% of Barroso's land has been slated for mining prospecting or licensing. In addition to the 15 existing licenses, the government has recently identified an area of 550 km^2 in Barroso, which is intended for a state auction. The two most significant and most developed lithium mining projects in this region are the proposed "Mina do Romano," located in Montalegre, whose concession contract is owned by LusoRecursos Portugal Lithium; and the proposed "Mina do Barroso," located in Boticas and owned by Savannah Resources Plc. through its wholly owned subsidiary Savannah Lithium Lda., which competes alongside Serbia, Finland, and Spain to be the largest open-pit lithium mine in Western Europe.

The "Mina do Barroso" project is only recently in the hands of Savannah Resources Plc. In 2006, a license for the exploration of feldspar and quartz mineral deposits in a total concession area of 120 hectares was issued. The contract was signed between the Portuguese State and Saibrais—Areias e Caulinos, SA. In 2010, the concession rights were transferred to Imerys Ceramics Portugal, SA.[31] In 2011, without any public consultation or notice to local authorities,[32] Imerys updated the Mining Plan to include the enlargement of the concession area to 542 hectares and the inclusion of lithium.[33] In 2016, the revised Mining Plan was approved by DGEG, with an addendum to the concession agreement, expanding the concession area to 542 hectares and including lithium as a concession substance. In the following year, the concession was transferred to Savannah Lithium, Lda.[34] Savannah now anticipates the expansion of the mine to 594 hectares.

The "Grandão Deposit" would be 500 meters long, 450 m wide, and 50 m deep; the "Reservatório Deposit" would be 400 m long and 100 m deep; the "NOA Deposit" would be 200 m long and 50 m deep; and the "Pinheiro Deposit" has unknown dimensions. The Environmental Impact Assessment (EIA) anticipates the expansion of the total concession area to 594 ha.[35] Yet, the document produced by Minho University commissioned by Savannah predicts a total area of 680 ha.[36] The average lithium extraction from the mine is expected to approach 1,450,000 tons of lethiferous pegmatite per annum, during the anticipated eleven to twelve-year life of the mine, corresponding to a production of 175 kt/y of spodumene concentrate (6% Li_2O).[37] The project anticipates that 86% of the production will be exported, corresponding to an average annual value of €110.2 million.[38] Savannah[39] claims the potential mine would follow the "world's best environmental practices for the minerals production industry," and argues this project has "the potential to contribute over €1.2 billion to Portuguese gross national product over the life of the operation." Savannah[40] was subject to public consultation between April and July 2021, after having filed two EIA, which were declared "non-compliant" by the Portuguese Environment Agency (APA) in 2020. Currently, the APA is reviewing the outcomes of the public consultation phase and, it recently declared,[41] it will announce its decision sometime in May 2023.

The potential mining site would cross the Village Councils (*Juntas de Freguesia*) of Covas do Barroso and Couto de Dornelas, impacting the small villages of Dornelas, Covas do Barroso, Romainho, and Muro. If the concession area extension is permitted, some houses in Romainho will be located only 50 m away from the mining site.[42] The proximity to these small villages will greatly impact the landscape, and, consequently its inhabitants. These are agricultural villages dominated by livestock production and crops typical of mountainous regions. Animal production is the basis of these towns' agrarian economies, dominated by extensive breeding of cattle for beef, namely for the Indigenous breed of "vaca barrosã." These towns maintain a rural subsistence economy with very few surpluses and relatively low consumption levels compared to other regions in the country.[43] The Barroso region, as mentioned before, is recognized for being an important biodiversity hotspot in Portugal, and has a high potential of becoming an important biodiversity conservation unit, as it hosts important populations of Iberian endemic plant and animal species,[44] including endangered species.[45] Moreover, the high rainfall levels of this mountainous region make Barroso one of the areas in Portugal with the best water resources.[46] The locals preserve ancestral ways of working the land and treating animals. For these reasons in 2017, as mentioned above, the Barroso region was the first in Portugal classified as a "World Agricultural Heritage" site and added to the "Globally Important Agricultural Heritage Systems" by the Food and Agriculture Organization of the United Nations.[47] The FAO has recognized the "authenticity of the territory, the traditional way of working the land, treating livestock, and the communitarianism of its inhabitants."[48] Indeed, the vast majority of the land in Barroso is common, known as "baldios" or "terras baldias."[49] In Covas alone, according to the estimates of the Directive Board of the Baldios of Covas do Barroso, there are approximately 2000 hectares of "baldios."[50]

Despite also covering private land, Savannah's mining project would mostly be located on common property, the baldios.[51] The baldios are a type of property of a specifically communal nature, whose administration and ownership is the sole responsibility of the "compartes," the owners of the common land[52] (Law No. 75/2017 of 17 August). To make use of the baldios, the company would have to sign a lease agreement with the Assembly of Compartes for a period of 20 years, renewable for a maximum of 80 years (Law No. 75/2017 of 17 August; Article 36). If no agreement is reached between the company and the compartes, it is possible for the state to expropriate the baldios (Law No. 75/2017, 17 August; Article 41). The company, however, would need to apply for expropriation for "public utility" (Law No. 75/2017, 17 August; Article 41). If the State grants the "declaration of public utility," the "compartes" would thus lose their rights to manage and administer the baldios for a given time period (Law No. 75/2017 of 17 August). In recent years, changes have been made to the legal framework shaping mining projects, specifically the amendment of Law No. 54/2015 of 22 June (commonly known as the "mining law") through the Decree-Law No. 30/2021 of 7 May. Albeit existing since 2015, this law was never active. The new mining law amendment adds the concept of "green mining" and further reinforces the fact that the land needed for mining projects can be expropriated for reasons of "public utility." European "public utility declarations," as pointed out previously,[53] emerge as a prominent method of "bureaucratic land grabbing."

"[I]F THE COMPANY HAD SUCH GOOD INTENTIONS ..." THE ARRIVAL

Preparation leading to mining projects in Barroso began with geological exploration in the 1990s.[54] As Nancy, a local teacher from Covas who has migrated to London, recalls:

> I was 18 or 19, and a geologist came knocking at our door on a summer holiday. Mom and dad were home … He wanted to look at a particular rock, someone had directed him to us, so my mother and I went with him … He took a bit of rock that stood out. And I said: "So, what is it exactly?" He said: "It's white granite, and it's very rare and it's very good for ceramics" … I'm not a scientist, but, you know, I wasn't completely satisfied with that [answer]. But I left it there because I had no way of figuring out what exactly he was interested in or what he was looking for. This was in the late 90s. It was July or August 1996.[55]

Nancy goes on to explain how she has started following reports about "her little small town" from London. As the years went by, she understood that vested interest in her region's geological resources was growing. People in Barroso had been familiar with rock quarries, and, to a lesser extent, with wolframite mining (from the Borralha mines), but not with open-pit lithium mines. "We never had much information, transparency," explains a local villager, "We had never heard about this [lithium]" and company representatives would tell people: "It was far from dishes for ceramics."[56] Company representatives, research participants contend, told them the mining project was for ceramics and glassware, something they were relatively familiar with. Others did not hear about the mine until the "controversy over the boreholes," as Robert claims, explaining how the mining company arrived "really without anyone knowing."[57] Savannah[58] drilled "more than 45 drill holes" because of "the need for [an] increase in the concession area." The depth and size of the holes left by Savannah's lithium prospecting alerted residents; the impacts left by the prospecting are still visible on the ground and on Google Maps. "There was [an] exploration [contract] since 2006 and practically no exploitation," explains Simpson, a local farmer, "[so] people were like: they just want to do more prospecting, 'okay.'" Simpson, who is a member of the Village Council, narrates the arrival of the British company in 2017:

> The company first talked to the President of the Directive Board [of the Baldios], and the president let them [do prospecting];

> then, they talked to the President of the Village Council, and they let them. Meanwhile, they talked to some private landowners … not so many … seven or eight people … [because] not everyone has land in the places where they did the prospecting. The company said they were just doing some prospecting. When asked what it was for, they didn't even know what it was for. They said it was prospecting for lithium … Lithium, we had never even heard of lithium [laughs]. And, of course, they were saying they already had a license, a permit from the government. Of course, hearing about the government permit, people are like: "The government is above it all," right?! And then the company also implied that if people didn't accept it good willing, they would be forced to accept it … So people let them [do the prospecting surveys] …[59]

Prospecting was initially permitted free of charge, but this policy eventually changed. After internal discussion within the Village Council, it was agreed to charge the company a €500 fee for each prospection. The company attempted to extend the terms to allow prospecting for another year, but, in the end, people managed to get their payment without signing this contract.[60] The company's attempt at free prospecting raised concerns: "[T]he company arrived here, knowing perfectly well that people have difficult living conditions," explains Simpson. "So, if the company had such good intentions, why did they ask to prospect the lands for free?!" The company for the last three years, has been forbidden from entering the fields.[61]

There were, already at the time, people working for the company within Covas. These *local intermediaries* were instrumental to the mine's arrival and to securing the long-term presence of the company. There were at least two individuals who embraced the project. "Jennifer" was the first person in Covas to work for the mining companies. Having studied environmental engineering and accounting, when the companies took an interest in her hometown, she viewed it as an opportunity. This allowed her to remain close to home—where "I wanted to work"—because, being "in a small place and receiving the same salary [as I would in a city], I could have better [living] conditions,"[62] explains Jennifer.

Applying with Sabrais in 2006, Jennifer was eventually hired in 2007 as a "Community Relations Officer" and stayed working with the project—through various companies—until 2020. Describing her community relations work, she explains:

> I would say: "I'm coming here on behalf of the company, the company wants to come to your land to do this [prospecting] work. Do you authorize it? Under which terms do you authorize it?" And that was it. I never tried to influence people to stop doing this or that, I would only say what we were proposing and they would accept it or not. I would say it in a very clear way.

Jennifer, along with John, tried to run for election for the Directive Board of the Baldios while working for the company.[63] When this happened, anti-mining organizing was already growing stronger. A group of locals found a competing electoral candidate, eventually winning the elections and preventing "mining candidates" from entering public office.[64] Jennifer, eventually leaving the company "over a year ago," does not regret working for them, and remembers—later self-censoring any critical words—that "when I joined [the company], the dimensions of the project were not like these," referring to the expanded scale of the mine. Part of the community relations duties passed to John, another person from an influential family in Covas, who began land contracting for Savannah.

While some yielded uncritically to prospecting by the company, most residents in the region had little to no idea regarding the company's prospecting and intentions, or the size, type, and depth of the mine. Nancy delivered the news of the large-scale open-pit mine in 2017, who—aware of the geologists' interests in her village—noticed the project resurface publicly online in English.[65] This initiated efforts to organize for more information about the mining project. Eventually, four consultations were carried out[66]—organized either by the City Hall of Boticas (*Câmara Municipal*) or by the locals.

Very early on, people recall, Savannah organized a meeting in the Romainho Chapel with "this guy who only spoke English."[67] In this meeting, Susan recounts, people were asking the Savannah representative: "If people are against the project, if the population doesn't accept it, will you [Savannah] stop?" The representative said: "No." Another woman recounts: "[H]e said ... it [would be] okay to send us out of here!"[68] This shocked the majority of the residents, as the company was not only completely disregarding local concerns but also bluntly acknowledging that the project would continue, even if it meant expropriating the land from local residents. "I didn't understand very well how you could do a project against the will of the population," explains Simpson.

> And one thing I realized very early on was that our opinion would never count, our opinion would only count if we accepted [the project] with compensation ... If we opposed or were against it, they were going to completely ignore us because we were of no interest to them. And that's what happened![69]

It was at this point people realized what was encroaching on the region. This resulted in formal organizing against the project and the formation of the United in Defense of Covas do Barroso (Unidos em Defesa de Covas do Barroso (UCDB)) Association in December 2018.

CONCERNS, REACTIONS, AND LAND CONTRACTING

Savannah[70] promises to "[i]nvest the best technologies available to the industry, aiming to achieve the best quality standards and the best environmental performance, as well as making mining activity more attractive to young people of active age." These company pronouncements, however, do not change that most villagers' reactions to the proposed lithium mine are negative. "Here we are all against [the project]," explains a woman, "people live honestly, from agriculture, and have a healthy environment ... why would they want a lot of money? I find it sad."[71] John, working on "land acquisition" for Savannah "since the beginning of 2021," admits that "even if [Savannah] wants to explain [the project], it can't. Unless it's door-to-door, person to person."[72] People are "afraid" of the project, but "most people are not aware of what the project is going to be," contends John. "This is the thing," he continues, "you can be the best speaker ... but if you've got two or three [people] to destabilize the situation, they'll only hear what they want to hear."[73] While a logical concern for John, the counter-accusations against the mine were systematic and worthy of comprehensive answers.

Despite the company's persistent efforts, the great majority of villagers in Covas and its neighboring towns remain vocal about the negative impacts of the project. "They are going to destroy everything … we're going to be left with nothing. Everything … will end," explains a farmer.[74] Elaborating these concerns, Susan states:

> [I]t's going to destroy everything that we have … the environment, the water quality … it's going to destroy everything. Then, as it starts to destroy, we're also going to have to end up leaving here, we are going to have to separate ourselves from our family … from our friends … and I don't want that.[75]

The mine will "work day and night, 24 hours a day, for eleven years or more."[76] Savannah[77] attempts to counter these concerns by promoting a Landscape Recovery Plan (LRP) that will implement "visual barriers" with "earthen bund walls and tree-shrubs," preserve "vegetation in the surroundings areas not affected by mining," and implement recovery work when "phasing-out" the mine.

The opposition to Savannah's mining project emerges from four principal sources. First, locals witnessed mining in the region while growing up, specifically the Borralha wolframite mine (1902–1986), which they describe as "a ghost town" and a "widow village."[78] The village of Minas da Borralha was created purposely for the mining activity in 1902, and was left abandoned after 1986, when the mining operations ended. As one drives through Borralha today, all that is left are empty houses, shattered buildings, ruins of mining equipment, and visible environmental degradation. That is why locals from this region refer to it as a "ghost town." During its peak, the Borralha mines employed almost 2,000 (mostly male) workers, many of whom have died, directly or indirectly from impacts related to the mining,[79] hence the collective image of a "widow village." This experience shapes locals' expectations and concerns regarding mining. Albeit recognizing that the mining activity of Borralha generated employment opportunities at the time, residents largely tend to stress their past, present, and future socioecological costs, as well as the fact that contemporary mining operations are highly automated and demand highly qualified skilled labor, which would likely mean the locals would not be employed in the proposed mine.[80]

Second, people did their own research on the impacts of mining, comparing the claims and outcomes of companies operating open-pit mines. "Nobody believes it [the company's claims] because people go on the internet, they see mining in other countries," explains a farmer.[81] Environmental social impact research, moreover, suggests the validity of these concerns.[82] Third, villagers express how they believe the "green mining" labeling or the "energy transition" rhetoric is a fallacy, repeatedly proclaiming that "Barroso is green!" ("*verde é o Barroso!*"). Locals contend that climate mitigation policies should not rely on advancing the devastation of entire ecosystems, with some claiming that "lithium is just being used as a Trojan horse" to advance capitalist resource extractivism.[83] Finally, the company's failure to provide clear and transparent information and communicate with the local inhabitants, further undermines their credibility. "They had never talked to the population!" explains the Boticas Mayor, Fernando Queiroga. When the company finally made a public consultation, on Queiroga's request, the experience was unsatisfactory. "That meeting made people even more confused," explains the mayor, "I myself, who had some information, left [the meeting] more confused because they put a technician to explain ... with very complicated, very technical terms!"[84]

Water remains a notable concern. "It all gets contaminated, because there is a river and the water—for now—is clear and pure, [but] tomorrow it's going to get contaminated," explain two neighbors in agreement.[85] Savannah[86] presents three options for "[s]urface water division and control structures," but contemplates the possibility that "in case of water shortage, [the] water [will be] captured directly from the Covas River." Given the reality of open-pit mines, their intensive water use,[87] and tailing dam failures,[88] among other consequences,[89] residents are worried. Renowned Portuguese naturalist, Ernestino Maravalhas, reminds us of the birds, but more so of the Shining Macromia Dragonfly (*Macromia splendens*), which remains a high-priority species for the EU and its sensitivity to river and water table conditions is well documented. "The European Union is investing thousands of euros into protecting the Shining Macromia, in several places," explains Ernestino, "and, at the same time, they are allowing the destruction of the habitat of such a rare species" in Barroso.[90] The mine, Maravalhas explains, will destroy, but also fragment environments, eliminating environmental conditions that would allow insects to travel and reproduce. This directly threatens the dragonflies', and other species', genetic pool. The "pollinators or other insects that live close to the mine will be destroyed, they will not have a genetic influx"—reducing the breeding grounds of insects—explains Ernestino, stressing: "If you place the mine in this [Shining Macromia & Alcon Blue] habitat, it will be worse than a bomb."

Lithium mines notoriously absorb and pollute water tables. Savannah[91] "foresees a water need of 0.570 hm^3 of water for its first year of operation and 0.510 hm^3 for the remaining years of the mining operation." Presenting these figures in cubic hectometers (hm^3), Savannah attempts to conceal the mine's intensive water demand to the public, which converts into 570,000,000 liters the first year and 510,000,000 liters the remaining eleven years. This translates into 47,500,000 liters a month and 1583,333 liters a day, which is likely a conservative water-use estimate by the company. Renowned hydrologist Steven H. Emerman[92] has issued a report—commissioned by the UCDB—explaining that Savannah's project is "highly experimental" and does not comply with safety standards, especially when it comes to water management infrastructure for the waste mound and filtered tailings. Recognizing the high levels of biodiversity and rich ecosystems in Barroso, Ernestino ends by pleading: "Please, do not kill people and nature in Barroso. They are healthy as they can be and they do not deserve to be killed by the mining process."

The mine, Savannah contends, is offering social development, employment, the latest mining technologies, and reclamation schemes to minimize ecological impacts.[93] Benefits and mitigation actions, according to Savannah and their local representatives, will lead to rural prosperity and the repopulation of the region, which has witnessed a demographic decline since the 1980s. Offering to buy land above fair market value, Savannah seeks to convince residents to sell their land. As is similar in other sites,[94] land regularization, measurement, and, overall, territorial legibility to make the mining possible remain an issue. Land titles need updating within families, topography is difficult to identify and it does not always correspond to state tax records. This is because, according to John, the "land used to be divided by landmarks, or walls." Savannah employs landscape crews to create visibility over the vegetation, reveal landmarks, and "then be able to survey" the land. Savannah, accordingly, presents this work as a public service, though they have reportedly marked fields wrongly.[95] According to a farmer, Savannah claims to have "cleared the land, we measured it, just to know and then they try to make promise of purchase of sale contracts."[96] Spatial legibility for resource control resonates with state territorialization strategies,[97] which is complemented by long-term land rental strategies typical of land control strategies to acquire a large portion of the communal land.[98] Savannah is offering €2 per square meter of uncultivated land, €2.5 for cultivated lands and an additional €2,500 bonus for water springs and buildings on the land, yet the latter depends on the "surveyor, and he gives a price." Land contracting currently includes a survey team: two people working in land acquisition and a landscape company—Landfound—comprised of three men. The company's tactics, so far, have resulted in few gains: after significant efforts, only less than half a dozen people have sold their land.[99]

Rural prosperity is linked to land rents, jobs, social development funds, and, since the new mining Law No. 30/2021 (May 2021), increased rents for municipalities. "Decree-Law No. 30/2021 now establishes that a part of the royalties (between one-third and one-half) is to be paid to the municipalities in which the exploitation takes place," explain Aroso and Magalhães,[100] the rest goes to the Portuguese state.[101] Savannah,[102] meanwhile, claims the mine will "create approximately 215 direct jobs and between 500 and 600 indirect jobs to support the project." Queiroga reported that in "June last year [2021], they started distributing some pamphlets asking for 200 employees for the mine without having the Environmental Impact Assessment approved yet!"[103] Savannah representatives, however, do not try to support these employment numbers, instead remaining ambiguous on the matter.[104] The greatest motivation for Jennifer and John to work for Savannah, besides the high-paying salary, was rural development. They wanted to prevent Covas from "becoming deserted," which relates to dependence on EU agricultural subsidies and stagnating meat and agricultural prices.[105] Covas residents, however, remain unconvinced by the company's social and environmental claims, which are rather extensive.[106] People are refusing the project, viewing the mine as destroying their environment, social fabric and way of life or, at the least, being a bad exchange. "Everyone likes to earn money and have a good salary, but to have that good salary, you're paying a high price. You have a good salary, but you are going to be living next door to an open-pit mine," Simpson reminds us.[107]

COMPANY ENTRENCHMENT, GREEN PROPAGANDA, AND THE COERCIVE INFLUENCE OF STATE POWER

When awareness spread about the company's intentions, in 2017, local populations from Barroso organized and established opposition to mining. From 2019 onwards, mobilizations have extended to the urban centers, with urban populations mobilizing themselves in solidarity. For the past five years, there have been several anti-mining mobilizations in major cities (Coimbra, Lisbon and Porto,) and in affected towns, including anti-mining protest camps, demonstrations, creating blockades against "pro-mining" governmental authorities, and numerous acts of direct action, claimed and unclaimed, such as barricades, graffiti and vandalism against governmental ministries, company and supporting contractors (e.g., University of Porto; see Figure 6.2). This includes one of the APA's offices being vandalized with "Minas Não" (No to Mines) in Porto, and a recent blockade to Savannah's information point in Boticas. Farmers, concerned citizens, climate movements, autonomists and anarchists are organizing to defend ecosystems from mining. Some leftwing political parties have slowly started integrating this issue into their discourse during the most recent electoral campaign. This has generated a response and adaptation from Savannah, and other mining companies in the region. Within this response, we can identify numerous social technologies of pacification, related to the professionalization of company activities. This section will discuss social warfare techniques designed to promote, legitimize and enforce green extractivism in this region.

Figure 6.2 University of Porto, spray painted with "Minas Não" (No Mines) and "Capitalítio," a play on words indicating capitalist lithium. Source: Twitter, MinasNao2.

Governmental and municipal power's coercive influence

Governmental authorities on the national, regional, and local levels—with the exception of the Village Council (*Junta de Freg-*

uesia) of Covas and the City Council (*Câmara Municipal*) of Boticas—are understood by our research participants as colluding with mining companies. On the national level, as discussed in the introduction, this relates to the locals' perception of the government selling off this region to mining companies without prior consultation of regional authorities and populations. The premature approval of several mining projects by the national government gives little hope to people, who repeatedly voice that they cannot count on or trust the government to act in their interest.[108] Remembering how the initial mining project was permitted as a 120-ha mine for quartz and phosphate in the early 2000s, the Boticas Mayor explains:

> This was authorized. From then on, the area was extended to 590 hectares without authorization from the Boticas City Hall (*Câmara Municipal*), or the APA, or the Ministry of Agriculture, or the CCDR [Commission for the Coordination of Regional Development], or the ICNF [Institute for Conservation of Nature and Forestry]. There was no authorization from anyone! There is an entity here called DGEG that authorized it without giving anyone a heads-up! So, what is going on?!

This represents the coercive and deceptive reality of bureaucratic land grabbing, which regional authorities experience. This dismay at the approval process echoes through the Barroso region. "[T]he government comes along and says it has nothing to do with [approving mines]," explains a local activist, "when the Directorate-General of Energy and Geology [DGEG] is directly dependent on them!"[109] The international promotion of mining tenders in Portugal through promotional videos and booths at the 2018 PDAC Convention, mentioned in the introduction, is rather damaging in this regard. This manifests as resentment toward the Portuguese government: "[t]he biggest culprit in all this is our government, who doesn't stand up for us," explains a local beekeeper.[110] The recent delay of the Savannah project related to the Environmental Impact Assessment might represent stalling, or regrouping, efforts by the Portuguese government, which nevertheless remains an obstacle for the company.[111] This extends to the denial of access to documents and information about the mining projects to residents and authorities. For example, the Portuguese Environmental Agency (APA) is currently under investigation by the Aarhus Convention Compliance Committee. At issue is the Agency's refusal to disclose information and denying access to documents related to the Environmental Impact Assessment of the Savannah Mine. Although still under investigation, this process is symptomatic of the way in which citizen's participation has been highly conditioned.

This national compliance and approval of mining exists at the municipal level in the Barroso region. This is especially the case in Village Councils within the jurisdiction of the Montalegre City Hall—whose mayor, Orlando Alves, is understood to be in favor of the mines. The organization of municipal power in Montalegre "makes people afraid to fight, even though they don't want the mines. They want to protect their lives and they have fears," explains a local activist.[112] This concern was repeatedly voiced throughout our interviews and informal conversations. "There are many people who depend directly or indirectly on the City Hall," explains a local ecologist.[113] "Therefore, many people, even if they are against the mines, will not say [it] for fear of consequences and retaliations [from elected officials]."

According to our research participants, some people living and working in Village Councils or City Halls with pro-mine candidates "don't say anything because they are afraid of losing their jobs,"[114] explains Simpson. He goes on to say: "A lot of the jobs that people have are dependent on the City Hall ... For example, people who work on construction sites need permits from City Hall." Research participants relate this to the political heritage of having lived under a "dictatorship until 40 years ago,"[115] which, an activist explains, has cultivated a "spirit of subservience, dependence, silence, of not talking about things, being vassals. It is the result of fascism and it is still inside people's heads today."[116] In matters of megaprojects, this translates into defeatism: "People always end up losing, they [feel like they] can never win."[117]

The challenge for pro-mining authorities is to stay in power with high levels of open and silent opposition. This has resulted in suspicion of irregular political decision-making processes. According to a local anti-mining organizer from Montalegre, "[i]t is mandatory that the [mining] subject goes to the assemblies of the City Hall" but the mayor "never took the issue to the City Hall meetings, and he always signed alone. So, we never knew—even though we read the [transcript] minutes of these meetings—what was happening because he always signed everything in his office, hidden, nobody knew anything."[118] Village Councils (*Juntas de Freguesia*) in favor of mining are withholding information, marginalizing critics, and creating administrative difficulties for mining opponents, such as not handing out certified proof of family care or other work and state-related notaries, according to our research participants.[119] Furthermore, a Village Council under the jurisdiction of the Montalegre City Hall is reported to have lent municipal office space to a mining company, according to our interviewees.[120] Currently, the Montalegre City Hall is under numerous corruption investigations by the Portuguese state.[121] Twenty-five defendants are being investigated for suspicions of corruption, prevarication and abuse of power. At stake are direct adjustments and payment orders made to companies owned by Orlando Alves' relatives.[122] Recently, in October 2022, Orlando Alves was arrested and indicted for suspicions of criminal association, prevarication, undue receipt of payments, document forgery, and abuse of power.[123]

Moreover, suspicions have been raised about the ties between the executive power (at the Montalegre Municipality) and mining companies' representatives.[124] Additionally, the use of votes from working migrant residents, that is, people from Barroso who have migrated abroad, especially to France, are instrumental to electoral strategies in the region. The migrants "are outside, don't have much connection to the land, and don't understand the fight we are making here [in Portugal],"[125] explains the beekeeper. Some of these migrants, research participants' claim, are used to affirm political power in different Village Councils in the Barroso region. "When it's voting time, they come here [Portugal] to vote, but they don't know what is going on here. They come in buses [from France] to vote," explains Julie. Some research participants claim migrants are not only used to maintain municipal power, but that they are strategically brought back with buses paid for by their Village Council and potentially, as we are told, co-financed by mining companies.

This allegation was discussed with a lawyer who, supposedly told them they had a case, but it would cost them a lot of money, which prevented research participants from perusing the case. Suspicions of electoral fraud, moreover, are already being investigated by the Public Ministry (Ministério Público), which has opened an inquiry into the case of suspected electoral fraud in Montalegre. Just as research participants explained, during elections in Montalegre, hundreds of migrants were received by a mayor at the Porto Airport and transported to their municipality to vote.[126] The maintenance of municipal power in the Barroso region (especially in the municipality of Montalegre) demonstrates the social engineering and relationship of force employed to stabilize political terrain to permit open-pit lithium mining extractivism.

Soft technologies of social pacification

The ecological or "green" claims promoting and justifying the mines in the region remain another inflammatory issue. Savannah employs a series of "soft" technologies of social pacification, namely making social development promises, organizing local support, employing public relations campaigns, utilizing scientific research to legitimize their claims, employing environmentally friendly rhetoric, and developing infrastructures to support its operation. By stressing deleterious local conditions (e.g., policy-driven rural depopulation, agricultural stagnation), the company uses land rents/sales, employment, social development funds, and mining revenue fees as key selling points for lithium mining in Barroso. Building on the rhetorical device of green mining, state authorities relate—if not equate—lithium mining with quarries in the region.[127] This seeks to downplay the damage and water use of lithium mines and make lithium mining relatable to the public.

Political technologies of pacification have progressively escalated since Savannah's arrival. The company developed public infrastructures to promote the mine—namely information centers, or kiosks: one located in the center of Covas in 2017, and, in 2022, a new one has opened in Boticas. The information point at Covas is a small room (see Figure 6.3), with a large-scale model of the proposed mine, tailing dam and processing facilities, a shelf displaying samples of products lithium can produce, and posters with the history of the mining project and the company's Landscape Recovery Plan. The center promotes the mine, serving as a point of contact for information, work, and land contracting. This infrastructure, likewise, includes Savannah billboards in the region and sponsoring a "rally runner" racecar in Boticas. The new info-point in Boticas, on the other hand, is located on the main city square, in front of the City Hall, in a modern and renovated building. Its contents are similar to the ones in Covas, showing a representative model of the mining project; several images; geological material; and several documents, including the EIA and the monthly newsletters sent to the community.

Figure 6.3 Savannah's Covas Information Point. "Savannah: Enabling Europe's Energy Transition." Source: Author, 2022.

This infrastructural influence links with symbols of wealth. People working for the company own notoriously expensive trucks (and display them, by driving them around) and overly modern houses when compared to the typical rural architecture of most houses in Barroso. Referring to a Savannah land contractor, Julie explains how "he comes around with people with high-cylinder cars, so you can see they are people with possibilities." This, she contends, "is part of the company's tactics to make locals associate the fact of working for the company with the possibility (at last!) of being able to buy luxurious material goods: 'Oh, with the money I'll buy that car!'"[128] This speaks to cultivating the modernist dreams of development, which the post-development school[129] has deconstructed at length.

Twenty-five kilometers away from Covas do Barroso, in the municipality of Montalegre, LusoRecursos (the mining company aiming to open the "Romano Mine") "has five electric Jaguars: the owner [of the company], some friends, including some of the local authorities in Montalegre all have five white Jaguars," explains a local activist.[130] The purpose, he continues, is to "create the idea that people are going to get rich, that there will be no shortage of jobs and that everyone is going to drive around in an electric Jaguar," meanwhile ignoring the consequences and realities of critical raw material extractivism. Complementing the modernist or developmental messages of consumer society, the mining companies (and their collaborators) seek to penetrate the "hearts," "minds" and desires by various direct and indirect means to accept critical raw material mining regardless of the socioecological consequences.

Public relations efforts extend even to offering the traditional Christmas Cake (*Bolo Rei*) to residents of Covas, Romainho and Muro. Savannah hired a baker from the region to distribute the cakes, but—as a local beekeeper recounts—"nobody accepted it" and people who "accepted it at the time, later went to give them back."[131] Savannah would also sponsor sports teams[132] and perform token infrastructural works. For example, a villager—who is reported to be unemployed and dependent on social subsidies[133]—needed help to repair the roof, and John, on behalf of Savannah, offered to complete the work, which the villager agreed to. A family member of the roof recipient explains this type of strategy intends "to win people over. This is manipulation. It's to buy time and buy people."[134] Likewise, Savannah reportedly harassed a local villager who was dependent on welfare subsidies and alcohol, promising him a new house alongside a housekeeping service. The villager had received the support of local anti-mining villagers and refused this offer, but this example shows how Savannah targets the most vulnerable people to progressively win the "hearts" and "minds" of a disagreeable population.[135]

In addition to going door-to-door in Covas and neighboring villages, Savannah also produces a monthly newsletter that promotes the mine and the company's civil performance. The newsletter, however, complements a wider media strategy, which sponsors numerous regional magazines or newspapers, as well as regional radio advertisements. Savannah "has been on the radio for a long time with advertisements," explains Susan:

> [Savannah would say that the mine] was going to be "clean mining," it wasn't going to affect the waters, it wasn't going to circulate through the villages. That is what they said in the radio ads. And then, to counter it, I called the radio to dedicate a song to all the people who were fighting against the mines: *Pelos caminhos de Portugal*.[136]

Radio advertisements, however, were not targeting Covas directly, but nearby towns. The radio and newspapers "were always outside," explains Aida Fernandes, the President of the Directive Board of the Baldios, "always outside, never local, and they always talked about us as a 'community.'"[137] Nik, from Mining Watch Portugal, recognizes this approach as "implementing a divide and conquer strategy on a regional level." Recognizing mining pacification efforts, Nik explains:

> A few weeks ago, they opened up an information center, like the one they have in Covas, in Boticas, in the regional municipality itself. The majority of the population does not live in Covas, the voting population is in Boticas and only people of Covas are the most noticeably against the mine. So, for the company, it is easier to get support from people living in the same municipality but living further away from the mining impacts. So, they engage in this strategy of financing sports teams, racing events, festivals and whatnot. And, now, they have their second information center in the city of Boticas itself just to steer public opinion about the mine.

In the media, a local ecologist claims,[138] people

> don't hear that the mines will destroy, will destroy jobs, and will pollute waters ... I am talking about the national level, but I am almost sure that more than 50% [of people at the national level] are in favor [of the mines] because the media has performed a lobotomy on people.

Overall, Savannah has attempted to enact a professionalization strategy to meet EU standards, engaging stakeholders, employing university researchers, and laying the foundation for the social engineering of extraction in the politically feasible avenues available. Academic institutions are a visible form of how this lithium extractive frontier is legitimized by a vast political and scientific complex. The Portuguese Mineral Resources Cluster Association, whose objective is "the promotion of knowledge and the sustainable economic enhancement of mineral resources," has as partners several Universities and Polytechnic Institutes alongside a number of mining companies, including Savannah and LusoRecursos. For example, the University of Aveiro, part of this cluster association, has signed a cooperation agreement with Savannah on lithium batteries. At the same time, researchers affiliated with this university (part of the cluster association) are often vocal in the media about the positive impacts—and the necessity—of lithium mining in Portugal, thus aiming to construct a favorable rhetoric—under the veil of science and assumed objectivity—to extractivism.

The "lobotomy" mentioned above is the social construction of mining as "green," "clean," and an expression of climate change mitigation. The government and companies "sell us that mines are to protect the environment and fight climate change," the local ecologist explains, "when in reality this is just another pretext to create profits and benefit big companies"—a "Trojan Horse" of extractivism. For him, lithium and electric vehicles are strongly understood as counterproductive to mitigating ecological crises. Governments "are completely blind, obsessed with climate change, and forget other things, like biodiversity, which is the basis of everything," says this land defender. To think, he continues, you can "solve this problem of climate change with lithium, by destroying nature and biodiversity, makes no sense at all." This relates to how carbon, the dominant measure of ecological catastrophe, enables climate reductionism,[139] which allows the rebranding of mining as "green" and "clean." These environmental claims, Aida asserts, "don't make sense to us, because it's not true." "This is greenwashing," she contends, that allows people in cities to "convince themselves that they are being good people by buying an electric car." People living and working near or around proposed mining sites experience this as absurd because their energy use and impacts are minimal, they maintain a Globally Important Agricultural Heritage system and, in the words of Aida, "have already been sacrificed, like the whole rural Portugal, to dams and wind farms."[140] The Boticas Mayor reminds us that Barroso is

> the region in the whole country where the most clean energy is produced: hydroelectric and wind energy. In fact, three dams are located in this region ... [and] [e]very year, the municipality of Boticas emits 9 tons of CO_2. If the mine is opened, it will produce 92,000 tons of CO_2 per year, for 12 years.[141]

The green credentials of the mine generate resentment and dismay from the majority of residents, while others who are seeking economic opportunities feel hopeful and refuse to ask questions. Given the dams, wind turbines, forests and rural living, Aida asks: "Who contributes more to decarbonization than we do?"

The final mechanism employed by Savannah is the threat of land expropriation. Multi-pronged pacification efforts combine money with the threat of land expropriation. To secure land, the company pays 10% upfront and says it will pay the other 90% once the government approves the Environmental Impact Assessment. This strategy is further combined with cleaning (and measuring) the land. John encourages people to gamble: sell their land now, because, if the government does not approve, they have already made some money; if the government approves, people will get much better land payouts than under the terms of expulsion. Put this way, John presents the deal as a "win–win" situation.

Opponents of the mine, however, interpret this as "deceit" and "ruining" the village.[142] "It's threatening people," explains Fernando

Queiroga. "It's threatening that either they will sell or else they will go for expropriation and then they won't get anything."[143] The threat of government expulsion greatly benefits Savannah's land contracting efforts. The door-to-door approach John discussed earlier, means speaking to everyone about this dilemma, seeking to push people to sign or, at the least, sow fear, hesitation, and doubt in the people opposing mining. People contend that Savannah is targeting "those people that are more malleable,"[144] notably the poorer demographics or the aging ones, who often express indifference about their future or the region's: "If they come [the company] … at my age …"[145] or "at my age, it doesn't make any difference."[146] As the beekeeper explains, "He's [John] trying to get into people's heads … scaring them, saying: 'You have to sell the land, If you don't sell it, then they're going to expropriate it.'" This is consistent with the first public meeting, when the company was explicit that it would take the land, with or without the consent of the people. The prospects of an open-pit lithium mine, and the land contracting process, is producing fear, uncertainty and social division. This is why, during an interview with her husband, Lisa blurts out in the background when discussing this situation: "This is just war, just war!"[147]

LITHIUM MINING IS NOT THE ANSWER

This lithium rush is being engineered at an international level with European policy proclamations, targets, and funds, which leads to intervention efforts by companies to organize the social conditions to extract lithium and other critical raw materials, as is the example of current efforts taking place in northern Portugal. Barroso residents have, at first, underestimated the reality of prospecting and surveying work that makes subsoil and surface resources legible for exploitation.[148] The town halls' trusting disposition, along with municipal authorities and individuals collaborating with companies, encourages mining projects in the area in the wake of dams and wind turbines.[149] Mines, like other megaproject interventions in the region, are branding themselves as "green," "clean," and exemplars of "sustainable development" and climate change mitigation, which residents reject as "fake green mine propaganda" and an "argument that is already not true."[150] The mine seeks to significantly degrade, if not destroy, an area with rich biodiversity, cultural heritage, and world-renowned agricultural practices, which includes important forest and water resources. These consequences are far from being unintended—rather, mining companies, in (more or less direct) cooperation with state authorities, are promoting a hyper-modernist worldview based on climate reductionism and techno-fixes as discussed in Chapter 1, thus seeking to dismantle what is left of low-impact agrarian and traditional lifeways. In this era of socioecological catastrophe, it seems more important than ever to preserve and expand the renowned socio-agro-ecological traditions of Barroso that its populations have been conserving and protecting for centuries.

The Savannah mine, as mentioned above, has finally been approved, yet Savannah has to meet strenuous conditionalities. This includes, as journalist Catarina Demony summarizes, "limiting the removal of vegetation from the project area, not taking water from a nearby river and carrying out landscaping once extraction has ended" as well as providing a "socioeconomic compensation package."[151] If Portuguese regulatory authorities are serious (which is highly unlikely), then disallowing water to be taken from the nearby rivers will present a costly, if not impossible, challenge for Savannah, considering the mine will need a conservative estimate of 1,583,333 liters of water a day. Whether people in Covas and beyond will engage in permanent ecological conflict against the companies remains up in the air. However, the Covas do Barroso case demonstrates, yet again, how an extractive company gains a foothold within a region. The company here, and across all the cases (see Chapters 2–5), began colonizing the region through supportive municipal agents and local collaborators, which literally *permitted* precise interventions into the social fabric of these rural communities—authorities and people let the company into the area. This further reveals, as Peter Gelderloos argues in a review of state formation and (anti-)colonial struggles,[152] how leaders emerge as a dangerous entry and leverage points for colonial authorities or, in this case, how companies can enter, influence and, eventually, control communities either directly or via proxy and indirect rule. Representatives in Covas and Boticas, thus far, have taken an oppositional stance, while other villages and towns have not. The company, however, engages in a practice of social warfare to chip away at, stretch and tear the remaining communal social fabrics, with the objective of weakening them, preventing opposition and, consequently, allowing access to the water, mineral, timber and human resources in the area.

The seemingly mundane and normalized efforts of the media, public relations campaigns, event sponsorships, information centers and modernized symbols of wealth, notably vehicles, are having severe psycho-social and, with prospecting, ecological consequences even before the mine has had the official stamp of approval. Southern Copper Peru (Chapter 4), like Savannah, also engaged in targeted media and public relations outside the region, attempting to rally support from people more distant from the socioecological impacts. Air, but more so, water degradation will still impact them, even if slowly and over a long period of time. Company efforts at organizing a Social License to Operate and employing social engineering techniques are profoundly subtle, but retain harmful consequences. Recognizing the efforts to organize land defense activities and the corresponding hardships, Nik reminds us:

> There have been hundreds of hours spent in knowing what happened with the mine ... There are phone calls. "Oh, the company is now entering there, this other village; they are walking on the mountain, they are now driving out with their jeep." Sometimes it is the company or someone else, but this constantly being on alert has a psychological, but also a physical toll in the end. There is a lot of suffering, even when there is no physical violence involved, when normally people think of police violence or state violence in classical cases.

Severe psycho-social harms emerging before mines or dams construction echoes findings from Reinert,[153] Velicu,[154] Batel, and Küpers.[155] The company, moreover, employs social warfare strategies designed to exhaust, demoralize, and defeat mining opponents in order to pacify the resistance and, consequently, access lithium (as well as other critical raw material resources). These insidious strategies, especially among the most concerned residents, are already causing emotional exhaustion and drain within people. This is reminiscent of oppositional residents in France (Chapter 5), who describe the efforts of infrastructure companies and the police as engaging in "a war of attrition" organized with the "logic to get people tired and disinterested in the cause." Company strategies, government complicity and policy incentives are driving the negative socioecological mining trajectory, which must urgently be considered by policy to the general public. This reconsideration, of course, must mean organizing solidarity, mutual aid and direct action.

European environmental policy, as discussed in Chapter 1, organizes climate reductionism, narrowly focusing on CO_2 emissions, to promote neoliberal environmentalism or green capitalist approaches to socioecological catastrophe. While Barroso also suffers from the mechanization of food and economic liberalization as a result of European agricultural policies, it remains a region to learn and build from—an area ripe for scaling up and improving agro-ecological activities and initiatives. Barroso, like other rural and Indigenous territories confronting mining and infrastructure companies (see Chapters 2, 3, 4, and 5), is not the problem, even if regional collaborators seem quick to "sellout" their ecosystems, culture and climate-stabilizing natural areas. While this is understandable considering the dominant socioecological pressures and European standards, the (hyper)modernist trajectory of material and energy-intensive infrastructure necessitating extractivism is a local, national, and global socioecological and climate threat. The shift toward post-developmental strategies and degrowth—revaluing and appreciating traditional and convivial lifeways—needs to spread by every means necessary, instead of being marginalized by mainstream environmentalism and eco-modernism. European environmental policy encouraging mass extractivism of critical raw material remains a fundamental obstruction to genuine socioecological sustainability. Lithium mining in Barroso is not a viable solution, let alone a climate change mitigation strategy.

Conclusion: fighting to win

This book has explored five substantial environmental conflicts, which encompass other struggles that confront high-voltage power lines and wind turbines invading numerous towns in Oaxaca, France, Catalonia, and Spain. The previous chapters have examined how different energy infrastructure and mining projects have entered and taken hold within regions and communities, how people have resisted and how companies and governments work to undermine and negotiate with this opposition. While there were victories in Oaxaca, Germany, and Peru that remain partial or tenuous; France, Catalonia, and Spain witnessed the construction of an energy transformer, wind parks, and high-voltage power lines that are expanding capitalist energy markets in Europe. Meanwhile, the proposed lithium mine in Portugal has a looming presence, waiting on the Portuguese Environment Agency's decision, where people will either engage in permanent ecological conflict, capitulate to the company, or somewhere in between. What common features can be observed in this book and what lessons can be learned to "win" ecological struggles?

For starters, despite the different geographical locations, political histories, and cultures, common features emerge across all of the sites. The most daunting structural feature made clear in all these conflicts, is the existence, and dare say imposition, of state bureaucracies and capitalist operations over all these territories. Governmental actors, at many scales (and sometimes conflicting with each other), and national and transnational companies—along with NGOs, to mediate the conflicts—were all enduring features transposed across such diverse cultural and geographical terrain. In short, while social ecologies and landscapes are diverse, the technologies of socioecological and political imposition are rather uniform, even if they are enacted and negotiated differently within their respective locals.

In the cases in Chapters 2–6, governments, at least at the national level, remained complicit with energy infrastructure and extractive projects. Local municipalities, however, remained important sites of approval for megaprojects and, in some cases, contested them. Local mayors, or municipal agents, supporting megaprojects often led to their ousting either by force or voting. Political representatives, and municipal power were also instrumental points for approving and supporting extractive development projects. Companies would go to political leaders, land elites, and landowners, first and foremost, avoiding announcing projects and public consultation. Once public consultations arrive, they would use this space to promote the projects and offer token participation to people, avoiding to provide relevant socioecological information. Townhall events, or consultations, frequently retained little to no substance—at least until people took matters into their own hands.

We witnessed the rejection of consultations and the creation of popular consultations (*consulta popular*), which were complemented by organizing information outreach campaigns. Particularly glaring, at least in the Juchitán, Oaxaca, free, prior, and informed consent (FPIC) consultation, was the lack of detailed and definite information regarding the distribution of money, but more so the impacts on ecologies from wind turbine construction on animals, trees, water, and electromagnetic currents from high-tension power lines near or above homes or in and around fishing sites. Low-carbon infrastructures require numerous mines, from copper, nickel, bauxite, neodymium and more, smelting facilities, manufacturing plants, and large-scale transposition before the operational socioecological impacts begin: habitat clearance, soil compaction, absorbing ground water, killing birds, leaking oil (or other chemical resins in the case of solar panels), and degrading crops. Likewise, this is all powered by fossil fuels with a sprinkle of hydrological and wind resources. There is no separation between hydrocarbon and low-carbon or so-called renewable energy infrastructures. Mines and energy infrastructure, and the operation of these infrastructures, degrade ecosystems and generate systemic health impacts from this degradation. These matters are all contentious, and political, and relate to methods of knowledge (e.g., epistemology) and how bodies and ecosystems are related to and understood by people (e.g., ontology). How to farm lands and fishing sites will be changed from wind energy development retains particularly high stakes for local approval and an entire industry. Declining fishing populations, degradation of soils, and cultural erosion create severe, if not life-threatening, socioecological stresses that force people into extractive labor, and national and international migration patterns. This type of information, as well as the latest science and critique, is completely necessary for any local decision-making. Unfortunately, obstructing popular knowledge, participation, and socioecological and economic knowledge sharing are all reoccurring features within these sites. Local socioecological knowledge is frequently excluded and ignored, in favor of hired consulting firms.

Political leaders and elites such as land owners, priests, and entrepreneurs, are instrumental in facilitating extractive development within regions, providing a "foothold" and entry to begin social engineering localities, to accept projects and begin a process of land acquisition, habitat degradation, and/or destruction. The distribution of money, of course, is instrumental to shaping and engineering local mentalities and dispositions. "External" colonization, it should be remembered, always requires internal collaborators—people from those regions—to allow entry and, consequently, opportunities for colonizers to employ different political techniques, generate internal discord, and distribute confusing information regarding development projects.[1] Local authorities, while frequently "selling out" their town to the highest bidder in the name of social development—often forgetting about the costs of decommissioning infrastructure projects—are also instrumental in resisting extractive development. Municipal agencies, either taken by force in Oaxaca or vote against the projects in the Tambo Valley (Peru), Aveyron (France) and Aude (France), Girona (Catalonia), and Covas (Portugal) would enable resources, but even physical space, to organize, disseminate information and facilitate housing for outside supporters.

The support of political representatives, in some instances, had a unifying effect, spreading general concern through political constituencies and networks. Local governments are levers of social control and become sites of contestation and, thus, serve as important points in ecological conflicts. Mining and infrastructure companies are constantly seeking local collaborators and "mining candidates" to place in positions of power to facilitate their extractive operations, which comes with great monetary benefits for those leaders and their network—hence the appeal to "selling out." These efforts were thwarted in Covas do Barroso, while rectified by a takeover in Álvaro Obregón/Gui'Xhi' Ro and voting in Peru and multiple sites in France and Catalonia. Political representatives, while not an anarchist ideal, have the possibility to create important spaces and widen political struggles. If mayors and Village Councils are committed to people and the environment, all the better—if not, they have to go. Supportive municipal agents, however, can become targets and face intense political pressure and attacks from state and federal representatives and the companies themselves, a stress which is designed to have them betray their local constituency and the struggles they claim to support. This pressure should be anticipated by movements and mitigated to "win" ecological struggles, if even temporarily. Peru, moreover, is an example par excellence, leaders will say anything to get into positions of power and then, almost instantly, betray their constituency or popular movements. This, however, is an almost universal feature of politicians, North and South of the globe.

Civil Society groups and non-governmental organizations are also important actors, often initiating contestation and organizing resistance. This frequently includes public denunciation, petitions, filing lawsuits and engaging in protest and, later, direct action campaigns. Civil society groups make up diverse actors, yet stereotypically tend to differ from anarchist and autonomous militants—locally or from outside the region—which frequently generates clashes over political analysis and protest tactics. Civil society groups might just want to avoid mining and infrastructure projects in their backyards (e.g., NIMBY), while others might connect this to systemic features of statism and capitalism—viewing ecological struggles as having wider implications and being in solidarity with other struggles nationally and internationally protecting ecologies and fighting socioecological destruction. The connection between civil society groups and "militants," internal or external to a particular location of struggle, remains fundamentally important and—often—a completely frustrating process for both sides when people are narrowly focusing on personal interest and unaware of the protest strategies and tactics that can actually win struggles.

The short answer, to "how to win ecological struggles," is through a whole ecosystem of action against companies and the governments—or specific ministries—that support extractive development. Civil society groups, lawyers, anarchists, farmers, fishers, mayors, teachers—everyone—are needed to work toward the common goal of *stopping extractive development projects*, be it mining, factories, waste pits, or energy infrastructure. Then, people need to take responsibility for their dependence on and addiction to this system. This means getting serious about communal food, electricity and the socioecological cost of entertainment. Degrowth translates into turning green spaces into edible landscapes; sharing ideas, knowledge, and gardening techniques; proliferating forest gardens and farming landscapes to regenerate and not deplete the soil; refashioning educational systems to prioritize ecological knowledge and convivial technologies; reinvigorating communal self-organization; prioritizing ecological building techniques and materials; and relying on small-scale low-carbon infrastructures (wind, solar, hydro, biomass) that produce and consume electricity in the closest proximity possible.

There are many ways this can happen, through chaotic and anti-authoritarian organization, through autonomous self-governance and much more. Concrete examples include the Bookchin family's proposal for libertarian municipalism that seeks to revive direct democracy and collectively organize political futures through local municipalities.[2] This proposal has been taken up and expanded in Rojava,[3] while the municipal tradition remains strong in Catalonia.[4] Related to libertarian municipalism is Kirkpatrick Sale's bioregional proposal,[5] but more so numerous Indigenous traditions of self-organization that remain extensive. Every Indigenous nation retains a method of socioecological self-governance, which inspires "communing" practices (and research),[6] infuses Zapatista autonomous self-organization,[7] and in Bolivia and Ecuador led to the popular application of socio-political systems of Buen Vivir ("life in harmony") and Sumak Kawsay ("living well").[8] While the integration of Buen Vivir into academia and national constitutions has generated concerning and assimilating results,[9] there remain many proposals for and past examples of socioecological "solutions." Caution against government assimilation and authoritarian recuperation is warranted, as Buen Vivir has demonstrated,[10] yet while these past examples might be inspirational, more important is to recognize that the time is always ripe for locally produced experimentation, practice and refinement to create socioeconomically justified and liberated spaces. This requires a revival in imagination, self-knowledge and connection with animals, rivers, mountains, and trees—to take ourselves out of malls and computers and place ourselves back into habitats vibrant with rivers, mountains, trees, flora and fauna. In the last chapter of *The Solutions are Already Here*, titled "A Truly Different Future," Gelderloos performs an exercise in imagination by outlining a post-revolutionary transition to bring socioecological balance and regeneration back into the present.[11] In short, people are described as taking back their "power" by regaining control over the production of electricity, food, medicine and self-defense, which necessitates reprioritizing cultural values, becoming conscious about one's role within an ecosystem and understanding the impact present work and productivist oriented market lifestyles are having on the environment. Our "needs" and "wants" need to be destabilized by new experiences, desires and priorities, which begins with recognizing how our present desires and destructive socioecological lifestyles have been regulated and socially engineered by state formation, (neo)colonialism, and the institutions and practices (e.g., work, schools, police, and prisons) they engender. Permaculture, convivial engineering, Indigenous traditions, autonomous organization, anarchist mutual aid and combat will animate the degrowth movement, if one is to exist outside universities, NGOs and policy discussions. This is undoubtedly an enormous challenge whether working with statist institutions or taking an autonomous and anarchistic approach, but no matter this can be a dignified and pleasurable pathway. The hard fights, as the saying goes, are the only ones worth fighting.

If people want to "win," then everyone is needed. This entails operating like an ecosystem. There may be disagreements, misunderstandings and ignorance, but staying unified in the mission to damage and defeat extractive infrastructures and the administrations that enforce them, needs to remain the common goal; a common goal that happens under police helicopters, the proliferation of violent rumors, traffic stop harassment, teargas, imprisonment (short or long-term), beatings, murder, and threats of being branded a "terrorist" for seeking to protect water, trees, and—often ironically—just upholding the law itself that politicians and companies violate, or draft new laws to override, as they see fit. The cases in this book, and beyond, demonstrate this. Recognizing the need for "everything," means embracing petitions, picnics, protests, parties, arson, and sabotage, among others, which has been referred to as embracing a "diversity of tactics."[12] In the Stop Cop City struggle, fighting to defend the Weelaunee Forest in Atlanta (USA) from a police urban warfare center, "Jean" says "We've gone a step further" than a diversity of tactics, "We've created something that actually mimics the forest itself, this is *an ecosystem of tactics*." Jean continues:

> So it's not a bunch of things working against or in spite of each other, it's several tactics working in conjunction and in relation to each other. Everything from the Muskogee stomp dance to marches of preschoolers to leafleting the community old-school style, to windows being smashed, to people building tree houses in the forest and refusing to move. [It's] punk shows and dance parties and religious services and garden planting … and a lot of these things are difficult for some people to understand why they matter; why they're connected to each other, but it's important to understand that we have to reach every aspect of human society.[13]

Defending habits, and winning ecological struggles, requires engaging in permanent ecological conflict and building an ecosystem of tactics. Implicit here, and a reoccurring theme in the chapters, is the importance of the relationship between local people and outsiders coming to support struggles. This relationship remains fragile, and when a common cause emerges, different understandings about the world, lifestyles, fashions, and politics collide. To speak in cliches, farmers tend to get tired of anarchist squatter "hippies" occupying their villages, while anarchists can smell the opportunism and indifference of some locals toward extractive development, looking to find a new short-term income stream at the long-term expense of the environment. No matter, building this ecosystem of struggle, nurturing those relationships, and developing an ecosystem of tactics in permanent ecological conflict remains the goal to protect habits, water, soil, air, and (socio-cultural) traditions. This is a lot of work, yet people are not alone in this struggle to defend their habitats. There are many, and many more who will have to begin to take an interest as socioecological crises worsen—only technology, drugs, alcohol, coercion, and other methods of enchanting pacification can numb people to this reality.

As the chapters demonstrate, the role of companies, governments, or segments of them, and public and private police-military forces are dedicated to undermining opposition, and permanent ecological conflict and an ecosystem of tactics. Call this social warfare, or counterinsurgency, the goal is to divide and pacify opposition. While leaked documents, analysis, and ex-military mine private security guard, "Jim," can testify: "Without asymmetric warfare techniques the protesters can paralyze the mining operations, and in one day the company will lose millions." Extractive companies themselves gauge opposition by protest and disruption; if opposition, protest, and sabotage subside, this is interpreted as consent or, at the least, acquiesce. The fear of illegality and permanent conflict, compounded by media propaganda, turns political tactics into a contested site that frequently divides political movements. The "good/bad" protester dichotomy is unrolled by authorities, media, and church and speaks to middle-class morality. Effective tactics, incurring economic damage and delay, are "bad," while symbolic and economically insignificant tactics like "peaceful marches" are "good." However, there are no "good" or "bad" protesters, just tactical considerations and the necessity to remain unified as an ecosystem. The goal, it must be remembered, is the defeat of extractivisms and, in the short term, the destruction of companies or political regimes that support extractive operations. The defacement, damage, and arson of property is often not "poor conduct" but effective political tactics to indicate the rejection and unwelcomed actors in the region, incurring a cost for imposing themselves and engaging in untransparent maneuvers and negations within a region. Movements, and especially middle-class elements, often forget this, which stresses the need to understand the logic of (social) war imposed onto people and landscapes, to discipline bodies, and mentalities and acquire habitats with all the timber, minerals, water, air and soil that might entail. Disagreement, indifference and feuds exist within ecosystems, but it should never undermine the preservation of habitats. Critical solidarity remains a necessity for developing an ecosystem of action that needs little coordination, because the goals and objectives are clear: protect life, cherish water and resurge habitats against coercive governments and extractive companies.

In efforts to undermine permanent ecological conflict, companies employ social development, and especially medical clinics, to "win" people over or, at the least, create divisions and decrease the frequency of effective global protests. Companies, however, are trying to avoid exorbitant expenses and permanent conflict against their operations. Yet, we are witnessing the weaponizing of amenities, and health care, to acquire land, water, and minerals. This indicates the importance of "peoples medicine," convivial technologies, food and energy autonomy, or relative autonomy, to meet the needs of people and to break dependence on material and energy-intensive systems, and extractive actors. The postdevelopment school deserves an immense amount of credit, challenging the construct of "development," which became a synonym for consumerist capitalist economies, high-modernist technology, and economic growth as the emblem and aspiration of success.[14] The underside, however, was constructing traditional ecological knowledge, communal organization, and anti-capitalist lifeways as "underdevelopment." Development could not emerge without colonial, or statist, genocide and ecocide,[15] allowing the emergence of development discourses that, in turn, instilled dependency on high-income countries and legitimized the imposition and replication of (colonial) institutions. Development began a (neo)colonial process of inputting "need" to said institutions and emerging technologies; and was complimented to engineer addiction to consumerism.[16] This, of course, produced the onset of socioecological and climate crises and this book documents conflicts on the frontlines of this struggle.

The uncomfortable, but obvious fact, is that politicians, police, and entrepreneurs are the eco-terrorists. The exception, in this case, proves the rule. These occupations are a product of a larger capitalist social war reproduced by family, school and church, e.g. patriarchy. Social war extends beyond class and, despite peoples' failure to recognize it, this concerns everyone. Everyone—even mining company executives—has a short and long-term interest in improving air, soil, water, food, and relational qualities. This common interest extends to combating workaholism, drug addiction, and being held hostage by rent and wage labor, to name a few, that regiments our cultures, social ecologies, and environments to generate stress, misery, and pollution. We can make utopias, but we have to want to and, most of all, recognize the lifeways of everyone, which extends to animals, rivers, trees, and mountains. People, myself included, work against our own interests and what gives us life, which must begin changing today.

This raises some final notes to consider. First, Bakunin was right. "The sincerity of the revolutionaries was not at issue; rather, the very use of the state machine imposed an 'iron logic' that made state managers 'enemies' of the people. Activists do not change the state; the state changes them."[17] This alerts people to the psycho-social conditioning of bureaucracy and divisions of labor that emerges when "revolutionaries" are separated and confined into their own institutional or personal interests, separated from their ecologies—flowers, bees, trees, rivers, and so on. This, however, extends to all people born into and immersed in a social war over "resources"—human and nonhuman—to develop a nation, capitalist economy, and spread the relations of capital across the world. The purpose and force that is consuming people and the world has been theorized elsewhere,[18] but humans—especially industrial humans—need to change their relationships with the planet, institutional objectives, education, and purpose. Because, currently guided by statism and capitalism, socioecological and climate crises, along with technological development, are making far-fetched dystopias a reality—and *it does not have to be this way.*

Permanent ecological conflict, through an ecosystem of tactics are the way forward. Dictatorship is not the answer, as is being proposed by some Marxian intellectuals, nor are representative democracies reinforcing capitalism, statism, and ecological conquest. People have to be the change they want to see according to their needs, capabilities, and accomplices. When engaging in a process of permanent ecological conflict, it is important to remember that this entire statist and capitalist system—and all of its institutions—are designed, in military terms, to "demoralize" people—to imbue apathy, steal passion, believe the once real is impossible, and, most of all, steal joy from struggling for what we love. People must not become demoralized in their permanent ecological struggles. Political praxis and health praxis must be one. Social war is political, personal, and spiritual with the objective of subordinating people to state, capital, and religious patriarchy, which requires everyone and their political, ecological, and social knowledges to make the world they want to see. Movements must embrace a diversity of peoples, thoughts, lifeways, and actions, like an ecosystem, where all participants create their own habits and have their own pleasurable preferences and still live unified in (relative) solidarity. This is a process of healing, care, ecological interest, and self-defense. Ecological crisis is not inevitable and everyone can play a role in creating vital habitats, and improving food, water, air and relationships that will warm hearts and slowly reverse the existing socioecological climate crises.

Acknowledgments

While this work is single-authored, there is an immense amount of support from all over the world to make this book possible. Friends, family, lovers, and cat partnerships, which extends to Kung Fu teachers, acupuncturists, boxers, teaching administrators, and academic colleagues and collaborators. This work would not be possible without them, nor the people taking an active part in resistance who are fighting to defend their habitats from encroaching energy infrastructure and mines. This book is made possible because of the wide range of research participants who have shared their words to communicate their feelings, discontent, and dilemmas. I am grateful for their trust and hope that this published work honors their words, causes, collective self-reflection, and, more so, inspires people to act in their own habitats—wherever that may be.

The research for this book, again, had so much support and help from people. Professors, administrators, anarchists, farmers, fisherpersons, and upright citizens of so many ethnicities. I am grateful to "Mr. X" and "Banda," who have purposely remained nameless, which is a network of people and friends who supported my research work in Oaxaca for upwards of eight years. This includes professors, lawyers, moto taxi drivers, and devout Christians, whose support I remain grateful for, as this help has likely saved my life more than once. In Germany, Andrea Brock's invitation to join her in researching the Hambach coal mine, as well as the land defenders I knew, or became friends with, in the forest were the backbone of this research in Chapter 3. Andrea, moreover, continued to carry on researching the Hambach Forest, while producing numerous important works, academic and popular, on the topic of coal extraction in Germany. In Peru, the courage and determination of Carlo Eduardo Fernandez, who invited and worked with me in the Valle de Tambo region of southwest Peru, was the foundation of my research in Peru. Carlo's patience and commitment to working with me and doing intense and heartbreaking interviews in the area made Chapter 4 possible. Likewise, I am grateful for the invitations, discussions, and collaborations with Jean-Baptiste Vidalou, Louis Laratte, and numerous other individuals assisting and supporting the research in France, and more still in Catalonia and Spain. Finally, the chapter in Portugal would not have been possible without the translation work, patience, and charisma of Mariana Riquito, but also others eager to share and instigate a struggle against lithium mining.

Overall, I am grateful to all my friends who are fighting, struggling with the tumultuous reality of living a life in permanent conflict, negotiating questionable habits and living life in constant tension against oppressive powers and extractivism. This care extends to Sabo, my cat partner, for being kind, patient, and a cuddly little street tiger. Appreciation extends to all of the people who have cared, loved, and supported me during these ten years of living in and researching environmental conflicts—you know who you are. At this moment, gratitude also extends to two Serbian cats, one a Linx (Ris), for their intelligence, beauty, resolve, courage, and, of course, patient conversations, which seeded desire and self-reflection. As Malvina Reynolds once said:

> *God[s] bless the grass that grows thru the crack ... God bless the truth that fights toward the sun. They roll the lies over it and think that it is done. It moves through the ground and reaches for the air, And after a while it is growing everywhere, And God[s] bless the grass.*

Because, we, the grass, are living, and the concrete—wires, pipes, toxic materials—suffocates life. I hope my friends, and everyone, can rise above this suffocation, estrangement, and enchantment of concrete in the many forms it takes, cultivating high-quality lives and environments wherever they are.

Thank you everyone, you know who you are.

Notes

PROLOGUE

1. Dunlap, A. 2019. *Renewing Destruction: Wind Energy Development, Conflict and Resistance in a Latin American Context*. London: Rowman & Littlefield.
2. Espartambos refers to the Spartans from the film 300, which Tambo Valley residence identified with confronting police invasion. These individuals took a position of combative self-defense against the police, widely recognized for grabbing large sheets of tin and wood to use as shields to block the rubber bullets, bird shot and rocks coming from the police.
3. "What is the point of all of this [research]?" Disenfranchised with academia and academics, I had to really think was collecting these testimonies really worth it for the people I spoke with? Do I just walk about making people remember horror? While this will be discussed more, it was confirmed my interviews played a cathartic role for some. Otherwise reader, judge this book as to whether this work was worth it.
4. Sorry for the jargon word, this refers to discrimination to other world views and science(s) other than Western or modernist science.
5. Stoddard, Isak, Kevin Anderson, Stuart Capstick, Wim Carton, Joanna Depledge, Keri Facer, Clair Gough et al. 2021. "Three Decades of Climate Mitigation: Why Haven't We Bent the Global Emissions Curve?", *Annual Review of Environment and Resources* 46: 653–689.
6. Ramirez R. 2022. "A Man Who Died After Self-Immolating in Front of Supreme Court Was a Climate Activist". *CNN*. https://edition.cnn.com/2022/04/25/politics/supreme-court-climate-activist-dies-fire/index.html.

INTRODUCTION

1. See Porter, G. and N.K. Kakabadse. 2006. HRM perspectives on addiction to technology and work. *Journal of Management Development* 25(6): 535–560; Samaha M. and N.S. Hawi 2016. "Relationships Among Smartphone Addiction, Stress, Academic Performance, and Satisfaction with Life." *Computers in Human Behavior* 57: 321–325.
2. Glendinning, C. 1994. *My Name is Chellis, And I'm in Recovery from Western Civilization*. Boston: Shambhala; Alexander, B.K. 2008. *The Globalization of Addiction: A Study in Poverty of the Spirit*. New York: Oxford University Press.

3. Marya, Rupa, and Raj Patel. 2021. *Inflamed: Deep Medicine and the Anatomy of Injustice*. New York: Picador, p. 33.
4. Ibid., pp. 20–21.
5. Ibid., p. 156.
6. Ibid.
7. Dunlap, A. 2018. "End the 'Green' Delusions: Industrial-scale Renewable Energy is Fossil Fuel+." *Verso Blog*. www.versobooks.com/en-gb/blogs/news/3797-end-the-green-delusions-industrial-scale-renewable-energy-is-fossil-fuel; and the expanded version: Dunlap A. 2021 Does Renewable Energy Exist? Fossil Fuel+ Technologies and the Search for Renewable Energy. In: S. Batel and D.P. Rudolph (eds), *A Critical Approach to the Social Acceptance of Renewable Energy Infrastructures – Going Beyond Green Growth And Sustainability*. London: Palgrave, pp. 83–102.
8. Hickel, J. 2020. *Less is More: How Degrowth Will Save The World*. London: Random House, pp. 6–16; Gelderloos, P. 2022. *The Solutions are Already Here: Tactics for Ecological Revolution From Below*. London: Pluto.
9. Ibid.
10. Marya and Patel. *Inflamed*, p. 60.
11. Bonanno, Alfredo Maria. *The Anarchist Tension*. Elephant Editions, 1998.
12. Anonymous. 2022. "Between Storms: Anarchist Reflections of Solidarity with Wet'suwet'en Resistance." Montreal Contre-Information. https://actforfree.noblogs.org/post/2022/11/20/between-storms-anarchist-reflections-of-solidarity-with-wetsuweten-resistance/
13. Franta, B. 2021. Weaponizing economics: Big Oil, economic consultants, and climate policy delay. *Environmental Politics*: 1–21.
14. Dunlap, A. and A. Brock. 2022. *Enforcing Ecocide: Power, Police and Planetary Militarization*. Cham: Palgrave.
15. On the concept internal colonization, see Grosfoguel R. 2021. "The Global Contributions of Black Decolonial Marxism: A forward." In: S.J. Ndlovu-Gatsheni and M. Ndlovu (eds), *Marxism and Decolonization in the 21st Century: Living Theories and True Ideas*. New York: Routledge, pp. xii–xxiii.
16. Moore, J.W. 2016. *Anthropocene or Capitalocene?: Nature, History, and the Crisis of Capitalism*. Oakland: PM Press.
17. The Indivisible Committee (TIC). 2015. *To Our Friends*. South Pasadena: Semiotext(e), p. 32.
18. Moore, *Anthropocene or Capitalocene?*
19. Ferdinand, M. 2021. *Decolonial Ecology: Thinking from the Caribbean World*. Cambridge: Polity Press.
20. Justin, M. 2016. Accumulating Extinction: Planetary Catastrophism in the Necrocene. In: J.W. Moore (ed), *Anthropocene or Capitalocene? Nature, History, and the Crisis of Capitalism*. Oakland: PM Press, Oakland, p. 116.
21. Gelderloos, *The Solutions are Already Here*, p. 36.

22. Shiva, Vandana. 2002. *Staying Alive: Women, Ecology and Development*. London: Zed Books; Romanyshyn, Robert. 2002 [1989]. *Technology as Symptom and Dream*. London: Routledge.
23. See Mies, Maria. 2014 [1986]. *Patriarchy and Accumulation on a World Scale: Women in the International Division of Labour*. London: Zed Books; Shiva, *Staying Alive*.
24. Illich, Ivan. 1978. *Towards a History of Needs*. New York: Pantheon Books.
25. Fitzpatrick, Nick, Tim Parrique, and Inês Cosme. 2022. "Exploring Degrowth Policy Proposals: A Systematic Mapping with Thematic Synthesis." *Journal of Cleaner Production* 365: 1–19.
26. Hickel, J. 2020. "What Does Degrowth Mean? A Few Points Of Clarification." *Globalizations*: 1–7.
27. Burkhart, C., T. Nowshin, M. Schmelzer et al. 2022. "Who shut shit down? What degrowth can learn form other socio-ecological movements." In: N. Barlow, L. Regen, N. Cadiou et al. (eds), *Degrowth and Strategy: How to Bring About Social-Ecological Transformation*. London: Mayfly, p. 129.
28. See Asafu-Adjaye, J., L. Blomquist, S. Brand et al. 2015. "An Ecomodernist Manifesto." *Ecomodernism.org*. https://static1.squarespace.com/static/5515d-9f9e4b04d5c3198b7bb/t/552d37bbe4b07a7dd69fcdbb/1429026747046/An+Ecomodernist+Manifesto.pdf.
29. Bolger, Meadhbh, Diego Marin, Adrien Tofighi-Niaki, and Louelle Seelmann. 2021. *"Green Mining" Is a Myth: The Case for Cutting EU Resource Consumption*. Brussels: European Environmental Bureau Friends of the Earth Europe. https://friendsoftheearth.eu/wp-content/uploads/2021/09/Methodology-considerations-Annex-to-green-mining-is-a-myth.pdf; Leonida, Carly. 2019. "Making Scandinavian Mining Sustainable." *Engineering and Mining Journal* 220 (10): 24–29.
30. See Dunlap, Alexander. 2017. "Wind Energy: Toward a 'Sustainable Violence' in Oaxaca, Mexico." *NACLA* 49 (4): 483–88; Dunlap, Alexander. 2022. "The Self-Reinforcing Cycle of Ecological Degradation and Repression: Revealing the Ecological Cost of Policing and Militarization." In: Alexander Dunlap and Andrea Brock (eds), *Enforcing Ecocide: Power, Policing & Planetary Militarization*. Cham: Palgrave, pp. 153–176; Burke, Matthew J., and Nina L. Smolyar. 2022. "Demilitarize for a Just Transition." In *Enforcing Ecocide*. Cham: Springer, pp. 307–329.
31. EC. "The European Green Deal." Online: European Union, 2019. https://eur-lex.europa.eu/legal-content/EN/TXT/?qid=1576150542719&uri=COM%3A2019%3A640%3AFIN
32. See Asafu-Adjaye et al.; Vettese, Troy, and Drew Pendergrass. 2022. *Half-Earth Socialism: A Plan to Save the Future from Extinction, Climate Change and Pandemics*. New York: Verso Books.
33. Parrique T., J. Barth, F. Briens et al. 2019. "Decoupling Debunked: Evidence And Arguments Against Green Growth As A Sole Strategy For Sustainability." *European Environment Bureau (EEB)*. https://eeb.org/library/decoupling-

debunked/; Hickel, J. and G. Kallis. 2020. "Is Green Growth Possible?" *New Political Economy* 25 (4): 469–486; Tilsted, J.P., A. Bjørn, G. Majeau-Bettez et al. 2021. "Accounting Matters: Revisiting Claims of Decoupling and Genuine Green Growth in Nordic Countries." *Ecological Economics* 187 (1): 1–9; Vadén, T., V. Lähde, A. Majava et al. 2020. "Decoupling for Ecological Sustainability: A Categorisation and Review of Research Literature." *Environmental Science & Policy* 112: 236–244.

34. Hickel and Kallis 2020. "Is Green Growth Possible?" p. 82.
35. Vadén et al., "Decoupling for Ecological Sustainability," p. 243.
36. Barlow et al. *Degrowth and Strategy*, p. 23.
37. For more see Hickel, *Less is More*.
38. Fitzpatrick et al., "Exploring Degrowth Policy Proposals," p. 10.
39. This is an ill-fated assumption, prepare for the worse and hope for the best with degrowth.
40. This critique remains rather general from Orthodox, but some heterodox, Marxists. The careless attacks by Leigh Phillips and Matt Huber are textbook examples.
41. For strong Feminist critique see Nirmal, P. and D. Rocheleau. 2019. "Decolonizing Degrowth in the Post-Development Convergence: Questions, Experiences, and Proposals From Two Indigenous Territories." *Environment and Planning E: Nature and Space* 2 (3): 465–492; Rodríguez-Labajos, Beatriz, Ivonne Yánez, Patrick Bond, Lucie Greyl, Serah Munguti, Godwin Uyi Ojo, and Winfridus Overbeek. 2019. "Not So Natural an Alliance? Degrowth and Environmental Justice Movements in the Global South." *Ecological Economics* 157 (2019): 175–184; Dunlap, A. 2020. Recognizing the "De" in Degrowth. *Undisciplined Environments*, https://theanarchistlibrary.org/library/alexander-dunlap-recognizing-the-de-in-degrowth. The book, however, will soon delve into an anarchist critique of degrowth.
42. Barlow et al. *Degrowth and Strategy*, pp. 1–405.
43. Ibid., p. 50.
44. Aronoff, K., A. Battistoni, D.A. Cohen et al. 2019. *A Planet To Win: Why We Need a Green New Deal*. New York: Verso Books.
45. Chomsky, N., R. Pollin, and C. Polychroniou. 2020. *Climate Crisis and the Global Green New Deal: The Political Economy of Saving the Planet*. New York: Verso.
46. Sanders, B. 2019. The Green New Deal. *Bernie Sanders Campaign*. https://berniesanders.com/en/issues/green-new-deal/
47. Dunlap, A. 2019. Green New Deal Part II: Good, Bad & the Ugly. *Terra Nullius: Repossessing the Existent*. www.sum.uio.no/forskning/blogg/terra-nullius/green-new-deal-part-II-good-bad-and-the-ugly.html
48. Dunlap, A. and L. Laratte. 2022. "European Green Deal Necropolitics: Exploring 'Green' Energy Transition, Degrowth & Infrastructural Colonization." *Political Geography* 97 (1): 1–17.

49. Hund, Kirsten, Daniele La Porta, Thao P. Fabregas, Tim Laing, and John Drexhage. 2020. "Minerals for Climate Action: The Mineral Intensity of the Clean Energy Transition," p. xi. *The World Bank Group*. http://pubdocs.worldbank.org/en/961711588875536384/Minerals-for-Climate-Action-The-Mineral-Intensity-of-the-Clean-Energy-Transition.pdf
50. Dunlap and Laratte, "European Green Deal Necropolitics"; Dunlap, "Does Renewable Energy Exist?"
51. Zografos, C. and P. Robbins. 2020. "Green Sacrifice Zones, or Why a Green New Deal Cannot Ignore the Cost Shifts of Just Transitions." *One Earth* 3 (5): 543–546.
52. Dunlap and Laratte, "European Green Deal Necropolitics."
53. See for, example, Pollin, R. 2018. "De-Growth Vs A Green New Deal." *New Left Review* 112: 5–25; Burton, M. and P. Somerville. 2019. "Degrowth: A Defence." *New Left Review* 115: 95–104; Robbins, P. 2020. "Is Less More… Or is More Less? Scaling the Political Ecologies of the Future." *Political Geography* 76: 1–6; Gómez-Baggethun, E. 2020. "More is More: Scaling Political Ecology Within Limits To Growth." *Political Geography* 76: 1–12.
54. O'Connor, J. 1994. "Is Sustainable Capitalism Possible?" In: P. Allen (ed.), *Food for the Future: Conditions and Contradictions of Sustainability*. New York: Wiley-Interscience, p. 133.
55. Gelderloos, *The Solutions*, p. 160.
56. Stoddard, I., K. Anderson, S. Capstick et al. 2021. "Three Decades of Climate Mitigation: Why Haven't We Bent the Global Emissions Curve?" *Annual Review of Environment and Resources* 46: 653–689.
57. Franta, B. 2021. "Weaponizing Economics: Big Oil, Economic Consultants, and Climate Policy Delay." *Environmental Politics*: 1–21.
58. Huber, M. 2022. "Mish-Mash Ecologism." *New Left Review*. https://newleftreview.org/sidecar/posts/mish-mash-ecologism
59. Phillips, L. 2015. *Austerity Ecology & the Collapse-porn Addicts: A Defence Of Growth, Progress, Industry and Stuff*. New York: John Hunt Publishing.
60. Huber, "Mish-Mash Ecologism."
61. Heron, K. 2022. "The Great Unfettering." *New Left Review*. https://newleftreview.org/sidecar/posts/the-great-unfettering.
62. Martínez-Alier, J. 2002. *The Environmentalism of the Poor: A Study of Ecological Conflicts and Valuation*. Northampton: Edward Elgar.
63. Huber, "Mish-Mash Ecologism."
64. Malm, A. 2020. *Corona, Climate, Chronic Emergency: War Communism in the Twenty-First Century*. London: Verso Books, p. 167.
65. See, for example, Friends of Aron Baron. 2017. *Blood Stained: One Hundred Years of Leninist Counterrevolution*. Oakland: AK Press.
66. The "White Threat" refers to the loose confederation of anti-communist forces that fought the Bolsheviks during the Russian Civil War (1917–1923). The real counter-revolutionary forces, which the Bolsheviks would quickly call anyone socialist, anarchist or peasants that opposed their rule.

67. Ryan, J. 2012. *Lenin's Terror: The Ideological Origins of Early Soviet State Violence*. London: Routledge, p. 9.
68. See Malm, *Corona*.
69. This quote is from this documentary: www.youtube.com/watch?v=gfd-nbMd9BiE; see also Chomsky, N. 2005. *On Anarchism*. Oakland: AK Press.
70. Pateman, B. 2017. Cries in the Wilderness: Alexander Berkman and Russian Prisoner Aid. In: Friends of Aron Baron (ed), *Blood Stained: One Hundred Years of Leninist Counterrevolution*. Oakland: Ak Press, pp. 243–258.
71. Ryan, *Lenin's Terror*, p. 2.
72. Ibid.
73. Friends of Baron, *Bloodstained.*
74. Goldman, E. 2017 "My Disillusionment in Russia." In: Friends of Aron Baron (ed), *Blood Stained: One Hundred Years of Leninist Counterrevolution.* Oakland: AK Press, pp. 223–242.
75. For an extended discussion of the academic negligence and Leninist manipulations perpetrated by Malm see: Dunlap, Alexander. 2022. "Ecological Authoritarian Maneuvers: Leninist Delusions, Co-optation & Anarchist Love." *Forged Books*. https://forged.noblogs.org/files/2022/10/dunlap-ecological-authoritarian-maneuvers.pdf; Klokkeblomst. 2021. "Green Desperation Fuels Red Fascism: Andreas Malm's Authoritarian Leftist Agenda." *Return Fire*. https://usa.anarchistlibraries.net/library/klokkeblomst-green-desperation-fuels-red-fascism-return-fire; Ffitch, Madeline. 2022. "A Frontline Response to Andreas Malm." *Verso Blogs*. www.versobooks.com/blogs/5325-a-frontline-response-to-andreas-malm
76. Fabbri, L. 2017 [1922]. *Blood Stained: One Hundred Years of Lennist Counter-revolution*. In: Friends of Aron Baron (ed), *Blood Stained: One Hundred Years of Lennist Counterrevolution.* Oakland: AK Press, p. 20.
77. Bakunin, M.A., M. Bakunin, and B. Michael 1990 [1873]. *Statism and Anarchy.* Cambridge: Cambridge University Press, p. 177.
78. Ibid., pp. 178–179.
79. van der Walt, L. 2018. "Anarchism and Marxism." In: N. Jun (ed), *The Brill Companion to Anarchism and Philosophy.* Leiden: Brill, p. 535.
80. See Chomsky, *On Anarchism*; van der Walt, L. 2018 "Anarchism and Marxism."
81. This quote is from this documentary: www.youtube.com/watch?v=gfdnbMd9BiE; Chomsky, *On Anarchism*.
82. Bakunin, *Statism and Anarchy*, pp. 23–26.
83. Foucault, Michel. 1996 [1974]. "On Attica." In: Sylvère Lotringer (ed), *Foucault Live: Collected Interviews, 1961–1984*. New York: Semiotext(e), pp. 113–121.
84. Perlman, Fredy. 1985. *The Continuing Appeal of Nationalism*. Detroit: Red; Black, pp. 56 and 58.
85. Bakunin, M. 1872. "Letter to La Liberté." *The Anarchist Library*, p. 7. https://theanarchistlibrary.org/library/michail-bakunin-letter-to-la-liberte.

86. Ibid., p. 8.
87. Ibid., p. 7. This also might be the precursor of the recent concept of "insurgent universality."
88. Marx, K. 1926 [1875]. "Conspectus of Kakunin's Statism and Anarchy." *Works of Karl Marx 1874*. www.marxists.org/archive/marx/works/1874/04/bakunin-notes.htm
89. Césaire, A. 1972 [1950]. *Discourses on Colonialism*. New York: Monthly Review Press; Rodney, W. 1972 [2009]. *How Europe Underdeveloped Africa*. Washington DC: Howard University Press; Robinson, C.J. 2000 [1983]. *Black Marxism, Revised and Updated Third Edition: The Making of the Black Radical Tradition*. Chapel Hill: University of North Carolina; Roane, J.T. 2018. "Plotting the Black Commons." *Souls* 20 (3): 239–266; Ferdinand, M. 2021. *Decolonial Ecology: Thinking from the Caribbean World*. London: Polity.
90. John, H. 2010. *Crack Capitalism*. London: Pluto; Federici, S. 2009 [2004]. *Caliban and the Witch: Women, The Body and primitive Accumulation*. New York: Autonomedia; Barbagallo, C., N. Beuret, and D. Harvie. 2019. *Commoning with George Caffentzis and Silvia Federic*. London: Pluto; García-López, G.A., U. Lang, and N. Singh. 2021. "Commons, Commoning and Co-Becoming: Nurturing Life-in-Common and Post-Capitalist Futures (An Introduction to the Theme Issue)." *Environment and Planning E: Nature and Space* 4 (4): 1199–1216.
91. Debord, G. 1994 [1967]. *Society of the Spectacle*. New York: Zone Books; to a lesser degree (in terms of drawing on Marx): Vaneigem, R. 2012 [1967]. *The Revolution of Everyday Life*. Oakland: PM Press. See also Wark, M. 2015. *The Beach Beneath the Street: The Everyday Life and Glorious Times of the Situationist International*. New York: Verso Books.
92. Foucault, M. 1995 [1977]. *Discipline and Punish: The Birth of the Prison*. New York: Random House, p. 168.
93. Stirner, M. 2017 [1844]. *The Unique and Its Property*. Berkeley: Ardent Press, pp. 204–205.
94. Foucault, "On Attica." Also Polanyi, Karl. 2001 [1944]. *The Great Transformation: The Political and Economic Origins of Our Time*. Boston: Beacon Press.
95. Kass, H. 2022. "Food Anarchy and the State Monopoly on Hunger." *The Journal of Peasant Studies*: 1–20.
96. Foucault, M. 2003 [1997]. *"Society Must Be Defended": Lectures at the College De France 1975-1976*. New York: Picador, p. 62.
97. Gelderloos, *The Solutions*, p. 22.
98. Bonanno, A.M. 1998 [1996]. *The Anarchist Tension*. London: Elephant Editions; Bonanno, A.M. 1999. *Insurrectionalist Anarchism*. London: Elephant Editions; Weir, J. 2017. *Tame Words from a Wild Heart*. London: Active Distribution.

99. For more information on this, read the section "NGOs and social warfare" in Dunlap, A. and M. Correa-Arce. 2022. "'Murderous Energy' in Oaxaca, Mexico: Wind Factories, Territorial Struggle and Social Warfare." *Journal of Peasant Studies* 49 (2): 455–480. And also Gelderloos, *The Solutions are Already Here*.
100. López, C. 2014. "Mexican Prisons: 'That Which Stagnates Rots.'" *War on Society*. https://waronsociety.noblogs.org/?p=9219
101. Anonymous. 2011. "Letter to the Anarchist Galaxy." *The Anarchist Library*. https://theanarchistlibrary.org/library/anonymous-letter-to-the-anarchist-galaxy; Rodríguez, G. 2014. "Anarchist Dialouges: A Conversation with Gustavo Rodriguez." *Conspiración Acrata*. https://en-contrainfo.espivblogs.net/files/2014/04/dialoguesgustavorodriguez.pdf; Weir, J., A. Bonanno, Billy et al. 2015. Down the Maxi-Prison. *Elephant Editions*, Available at: http://actforfree.nostate.net/wp-content/uploads/2017/02/down-the-maxi-prison.a4.pdf; Cospito A., G. Rodriguez, and GPd Silva. 2020. "Incendiary Dialogues: For the Propagation of Anarchic Sedition." *Internactional Negra Ediciones*. https://actforfree.noblogs.org/files/2021/04/incendiary-dialogues-eng1.pdf
102. Gardenyes, J. 2012. "Social War, Anti-Social Tension." *The Anarchist Library*. https://theanarchistlibrary.org/library/distro-josep-gardenyes-social-war-antisocial-tension.
103. IDG. 2022. "Between Climate Chaos and Social War: An Interview with Peter Gelderloos." *Its Going Down (IDG)*. https://itsgoingdown.org/between-climate-chaos-and-social-war-an-interview-with-peter-gelderloos/
104. Foucault, *"Society Must Be Defended."*
105. Trocci, A. 2011 "For the Insurrection to Succeed, We Must First Destroy Ourselves." In: A. Vradis and D. Dalakoglou, (eds), *Revolt and Crisis in Greece: Between a Present Yet to Pass and a Future Still to Come*. Oakland: AK Press.
106. Gardenyes, "Social War," p. 10.
107. Ibid., p. 11.
108. Kilcullen, D. 2006. "Twenty-Eight Articles: Fundmentals of Company-Level Counterinsurgency." *IO Sphere* 2 (2): 29.
109. Ibid., p. 31.
110. FM3-24. 2014. "Insurgencies and Countering Insurgencies." *Department of the Army Headquarters*. http://fas.org/irp/doddir/army/fm3-24.pdf
111. Ibid., pp. 1-1-1-2.
112. Williams, K. 2007 [2004]. *Our Enemies in Blue: Police and Power in America*. Cambridge: South End Press.
113. Kilcullen, D. 2012. "Counterinsurgency: The State of a Controversial Art." In: P.B. Rich and I. Duyvesteyn (eds) *The Routledge Handbook of Insurgency and Counterinsurgency*. New York: Routledge, p. 130.
114. Copulsky, J.R. 2011. *Brand Resilience: Managing Risk and Recovery in a High-Speed World*. New York: Palgrave Macmillan.

115. Ingold, T. 2014. "That's Enough About Ethnography!" *HAU: Journal of Ethnographic Theory* 4 (1): 385.
116. Dunlap, A. 2019. *Renewing Destruction: Wind Energy Development, Conflict and Resistance in a Latin American Context*. London: Rowman & Littlefield, p. 10.
117. Nader, L. 1972. "Up the Anthropologist: Perspectives Gained From Studying Up." *US Department of Health*.
118. For more on this read Verweijen, J. and A. Dunlap. 2021. "The Evolving Techniques of Social Engineering, Land Control and Managing Protest Against Extractivism: Introducing Political (Re)Actions 'From Above.'" *Political Geography* 83: 1–9; Dunlap, A. 2018. "Book Review: The Anarchist Roots of Geography: Toward Spatial Emancipation by Simon Springer." *Human Geography* 11 (2): 62–64.
119. They are scattered across numerous notebooks, some maybe lost in numerous moves.

CHAPTER 1: THE SCIENCE OF MAINTAINING SOCIOECOLOGICAL CATASTROPHE

1. Franta, Benjamin. 2021. "Weaponizing Economics: Big Oil, Economic Consultants, and Climate Policy Delay." *Environmental Politics*: 1–21.
2. Stoddard, Isak, Kevin Anderson, Stuart Capstick, Wim Carton, Joanna Depledge, Keri Facer, Clair Gough, Frederic Hache, Claire Hoolohan, and Martin Hultman. 2021. "Three Decades of Climate Mitigation: Why Haven't We Bent the Global Emissions Curve?" *Annual Review of Environment and Resources* 46: 653–689.
3. Blaser, Mario. 2013. "Notes Toward a Political Ontology of 'Environmental' Conflicts." In: Lesley Green (ed), *Contested Ecologies: Dialogues in the South on Nature and Knowledge*. Cape Town: HSRC Press, pp. 13–27; Sullivan, Sian. 2017. "What's Ontology Got to Do With It? On Nature and Knowledge in a Political Ecology of the 'Green Economy.'" *Journal of Political Ecology* 24: 217–242.
4. Foucault, Michel. 1977 [1966]. *The Order of Things: An Archaeology of the Human Sciences*. London: Tavistock Publications.
5. Newfield, C., A. Alexandrova, and S. John. 2022. *Limits of the Numerical: The Abuses and Uses of Quantification*. Chicago: University of Chicago Press, p. 11.
6. Dunlap, Alexander. 2022. "The Self-Reinforcing Cycle of Ecological Degradation and Repression: Revealing the Ecological Cost of Policing and Militarization." In: *Enforcing Ecocide: Power, Policing & Planetary Militarization*, Alexander Dunlap and Andrea Brock (eds), 153–176. Cham: Palgrave.
7. Merchant, Carolyn. 1983. *The Death of Nature: Women, Ecology, and The Scientific Revolution*. New York: Harper & Row.

8. Shiva, Vandana. 2002 [1989]. *Staying Alive: Women, Ecology and Development*. London: Zed Books.
9. Carson, Rachel. 2002 [1962]. *Silent Spring*. New York: First Mariner.
10. Klinger, Julie Michelle. 2017. *Rare Earth Frontiers: From Terrestrial Subsoils to lunar Landscapes*. Ithaca: Cornell University Press.
11. EU. 2020. "Critical Raw Materials Resilience: Charting a Path Towards Greater Security and Sustainability." European Commission. https://eur-lex.europa.eu/legal-content/EN/TXT/PDF/?uri=CELEX:52020DC0474&from=EN
12. Meadows, Donella H., L. Dennis Meadows, Jtsrgen Randers, and William W. Behrens III. 1972. *The Limits to Growth: A Report for the Club Of Rome's Project on the Predicament of Mankind*. New York: Universe Books.
13. Søyland, L.H. 2021. "'The Population Question' in Degrowth and Post-Development." *Debates in Post-Development & Degrowth* 1: 112–129.
14. Education, Population. 2021. "Energy Use Comparison Infographic." *Population Education*. https://populationeducation.org/resource/energy-use-comparison/
15. Sullivan, Sian. 2009. "Green Capitalism, and the Cultural Poverty of Constructing Nature as Service Provider." *Radical Anthropology* 3 (1): 18–27.
16. WCED. 1987. *Our Common Future*. Oxford: Oxford University Press.
17. Ibid., p. 40.
18. See the documentary *Fairytales of Growth* (2020), minutes 12:07–20. www.youtube.com/watch?v=dQ4cpOKmde8
19. McAfee, Kathleen. 1999. "Selling Nature to Save it? Biodiversity and Green Developmentalism." *Environment and Planning D: Society and Space* 17 (2): 133–154.
20. There is an abundant literature here to read, consider these review articles: Dunlap, Alexander, and James Fairhead. 2014. "The Militarisation and Marketisation of Nature: An Alternative Lens to 'Climate-Conflict.'" *Geopolitics* 19 (4): 937–961; Verweijen, Judith, and Esther Marijnen. 2018. "The Counterinsurgency/Conservation Nexus." *The Journal of Peasant Studies* 45 (2): 300–320; Dunlap, Alexander, and Sian Sullivan. 2020. "A Faultline In Neoliberal Environmental Governance Scholarship? Or, Why Accumulation-By-Alienation Matters." *Environment and Planning E*, 3 (2): 552–579; Büscher, Bram, and Robert Fletcher. 2020. *The Conservation Revolution: Radical Ideas for Saving Nature Beyond the Anthropocene*. New York: Verso; Marijnen, Esther, Lotje De Vries, and Rosaleen Duffy. 2021. "Conservation in Violent Environments: Introduction to a Special Issue On The Political Ecology Of Conservation Amidst Violent Conflict." *Political Geography*: 1–3.
21. Parrique, Timothée, Jonathan Barth, François Briens, Christian Kerschner, Alejo Kraus-Polk, A Kuokkanen, and JH Spangenberg. 2019. "Decoupling Debunked: Evidence and Arguments against Green Growth as a Sole Strategy for Sustainability." *European Environment Bureau (EEB)*. https://eeb.org/library/decoupling-debunked/

22. Vadén, T., V. Lähde, A. Majava, et al. (2020) "Decoupling for Ecological Sustainability: A Categorisation and Review of Research Literature." *Environmental Science & Policy* 112: 243.
23. Merchant, *The Death of Nature: Women, Ecology, and The Scientific Revolution*; Shiva, *Staying Alive.*
24. Sullivan, "What's Ontology Got To Do With It?," pp. 217–242; Blaser, "Notes Toward a Political Ontology of 'Environmental' Conflicts."
25. Romanyshyn, Robert. 1989. *Technology as Symptom and Dream.* London: Routledge, p. 35.
26. Ibid., p. 74.
27. Ibid., p. 82.
28. Dunlap, Alexander. 2014b. "Permanent War: Grids, Boomerangs, and Counterinsurgency." *Anarchist Studies* 22 (2): 55–79.
29. See Merchant, *The Death of Nature: Women, Ecology, and The Scientific Revolution*; Shiva, *Staying Alive*; Plumwood, Val. 2002. *Feminism and the Mastery of Nature.* London: Routledge; Bonneuil, Christophe, and Jean-Baptiste Fressoz. 2016. *The Shock of the Anthropocene: The Earth, History and Us.* New York: Verso Books; Daggett, Cara. 2019. *The Birth of Energy: Fossil Fuels, Thermodynamics, and the Politics of Work.* Durham: Duke University Press.
30. Blaser, Notes Toward a Political Ontology of 'Environmental' Conflicts," p. 13.
31. Daggett, *The Birth of Energy*, p. 174.
32. Sullivan, "What's Ontology Got To Do With It?," p. 226.
33. Daggett, *The Birth of Energy.*
34. Ibid.
35. Clarke, Bruce. 2001. *Energy Forms: Allegory and Science in the Era of Classical Thermodynamics.* Ann Harbor: University of Michigan Press.
36. Daggett, discussant presentation at the book launch workshop, Batel, Susana, and David Rudolph. 2021. *A Critical Approach to the Social Acceptance of Renewable Energy Infrastructures: Going Beyond Green Growth and Sustainability.* Cham: Palgrave.
37. Ibid.
38. Lohmann, Larry. 2021. "Bioenergy, Thermodynamics and Inequalities." In: Maria Backhouse, Rosa Lehmann, Kristina Lorenzen, Malte Lühmann, Janina Puder, Fabricio Rodríguez, and Anne Tittor (eds), *Bioeconomy and Global Inequalities.* Cham: Palgrave Macmillan, pp. 85–103.
39. Daggett, *The Birth of Energy*, p. 160.
40. Daggett, *The Birth of Energy*, p. 3.
41. Malm, Andreas. 2016. *Fossil Capital: The Rise of Steam Power and the Roots of Global Warming.* London: Verso Books.
42. Daggett, *The Birth of Energy.*
43. Daggett, *The Birth of Energy*, p. 3.
44. Bonneuil and Fressoz, *The Shock of the Anthropocene*, p. 101.

45. Ibid., p. 102.
46. Ibid.
47. Bell, Shannon Elizabeth, Cara Daggett, and Christine Labuski. 2020. "Toward Feminist Energy Systems: Why Adding Women And Solar Panels Is Not Enough." *Energy Research & Social Science* 68: 1–13.
48. Walker, Brian H. 1992. "Biodiversity and Ecological Redundancy." *Conservation Biology* 6 (1): 18–23.
49. Escobar, Arturo. 2012 [1995]. *Encountering Development: the Making and Unmaking of the Third World*. Princeton: Princeton University Press, p. 203.
50. Escobar, *Encountering Development*, p. 203.
51. On existences, see Kröger, Markus. 2022. *Extractivisms, Existences and Extinctions: Monoculture Plantations and Amazon Deforestation*. Vol. 1. London: Routledge.
52. Kelly, Alice. 2013. "Property and Negotiation in Waza National Park." *Land Deal Politics Initiative (LDPI)*, Brighton, UK.
53. Dunlap and Fairhead, "The Militarisation and Marketisation of Nature," pp. 937–961; Büscher, Bram, and Robert Fletcher. 2018. "Under Pressure: Conceptualising Political Ecologies of Green Wars." *Conservation and Society* 16 (2): 105–113; Marijnen et al., "Conservation in Violent Environments," pp. 1–3.
54. Sullivan, Sian. 2006. "Elephant in the Room? Problematising 'New' (Neoliberal) Biodiversity Conservation." *Forum for Development Studies* 33 (1): 105–135.
55. Sullivan, Sian. 2010. "'Ecosystem Service Commodities'—A New Imperial Ecology? Implications for Animist Immanent Ecologies, with Deleuze and Guattari." *New Formations: A Journal of Culture/Theory/Politics* 69: 111–128.
56. Ibid., p. 116.
57. Verweijen and Marijnen, "The Counterinsurgency/Conservation Nexus"; Marijnen et al., "Conservation in Violent Environments."
58. Dowie, Mark. *Conservation Refugees: The Hundred-Year Conflict between Global Conservation and Native Peoples*. Cambridge: MIT Press, 2009.
59. Anderson, Kat. *Tending the Wild: Native American Knowledge and the Management of California's Natural Resources*. Berkeley: University of California Press, 2005; Fairhead, James, and Melissa Leach. 1996. *Misreading the African Landscape: Society and Ecology in a Forest-Savanna Mosaic*. Cambridge: Cambridge University Press; Erickson, Clark L. 2008. "Amazonia: The Historical Ecology of a Domesticated Landscape." In *The Handbook of South American Archaeology*. Cham: Springer, pp. 157–183.
60. Dunlap and Fairhead, "The Militarisation and Marketisation of Nature"; Mirumachi, Naho, Amiera Sawas, and Mark Workman. 2020. "Unveiling the Security Concerns of Low Carbon Development: Climate Security Analysis of the Undesirable And Unintended Effects of Mitigation And Adaptation." *Climate and Development* 12 (2): 97–109.

61. Le Billon, Philippe. 2021. "Crisis Conservation and Green Extraction: Biodiversity Offsets as Spaces of Double Exception." *Journal of Political Ecology* 28 (1): 186.
62. Ibid.
63. Bolger, Meadhbh, Diego Marin, Adrien Tofighi-Niaki, and Louelle Seelmann. 2021. "'Green Mining' is a Myth: The Case for Cutting EU Resource Consumption." *European Environmental Bureau Friends of the Earth Europe.* https://friendsoftheearth.eu/wp-content/uploads/2021/09/Methodology-considerations-Annex-to-green-mining-is-a-myth.pdf
64. Hale, Charles R. 2002. "Does Multiculturalism Menace? Governance, Cultural Rights and the Politics of Identity in Guatemala." *Journal of Latin American Studies* 34 (3): 489. See also Ulloa, Astrid. 2013 [2005]. *The Ecological Native: Indigenous Peoples' Movements and Eco-Governmentality in Columbia.* London: Routledge.
65. Escobar, *Encountering Development*, p. 123.
66. Boukherroub, Tasseda, Yann Bouchery, Charles J. Corbett, Jan C. Fransoo, and Tarkan Tan. 2017. "Carbon Footprinting in Supply Chains." In: Yann Bouchery, Charles J. Corbett, Jan C. Fransoo and Tarkan Tan (eds), *Sustainable Supply Chains*, , 43–64. Switzerland: Springer.
67. Ibid., p. 46.
68. Ibid., pp. 51–52.
69. Ibid.
70. Ibid.
71. Bouchery, Yann, Charles J. Corbett, Jan C. Fransoo, and Tarkan Tan. 2017. *Sustainable Supply Chains: A Research-Based Textbook on Operations and Strategy.* Vol. 4. New York: Springer, p. 6.
72. Bolger et al., "'Green Mining' is a Myth."
73. Requena-i-Mora, Marina, and Dan Brockington. 2021. "Seeing Environmental Injustices: The Mechanics, Devices and Assumptions of Environmental Sustainability Indices and Indicators." *Journal of Political Ecology* 28 (1), p. 20.
74. Ibid., p. 21.
75. Ibid.
76. Dunlap, Alexander, and Diego Marin. 2022. "Comparing Coal and 'Transition Materials'? Overlooking Complexity, Flattening Reality and Ignoring Capitalism." *Energy Research & Social Science* 89: 1–9.
77. Gelderloos, Peter. 2022. *The Solutions are Already Here: Tactics for Ecological Revolution From Below.* London: Pluto, p. 38.
78. Dunlap, Alexander. 2015b. "The Green Economy as a Continuation of War by Other Means." *Capitalism, Nature, Socialism (CNS) Web.* www.cnsjournal.org/the-green-economy-as-a-continuation-of-war-by-other-means/ and Dunlap, Alexander. 2022. "The Self-Reinforcing Cycle of Ecological Degradation and Repression: Revealing the Ecological Cost of Policing and Militarization." In: Alexander Dunlap and Andrea Brock (eds), *Enforcing*

Ecocide: Power, Policing & Planetary Militarization. Cham: Palgrave, pp. 153–176.

79. Dalby, Simon. 2020. *Anthropocene Geopolitics: Globalization, Security, Sustainability*. Ottawa: University of Ottawa Press.
80. Junghans, Trenholme. 2022. "The Limits of the Numerical: Rare Diseases and the Seductions of Qualification." In: Christopher Newfield, Anna Alexandrova, and Stephen John (eds), *Limits of the Numerical: The Abuses and Uses of Quantification*. Chicago: University of Chicago Press, p. 93.

CHAPTER 2: GRABBING ISTMEÑO WIND

1. Elliott, D., M. Schwartz, G. Scott, S. Haymes, D. Heimiller, and R. George. 2003. "Wind Energy Resource Atlas of Oaxaca." National Renewable Energy Laboratory (NREL), p. iv. http://www.nrel.gov/wind/pdfs/34519.pdf
2. IFC. 2014. "Investments for a Windy Harvest: IFC Support of the Mexican Wind Sector Drives Results." International Finance Corporation, *World Bank Group*, p. 1. https://www.ifc.org/wps/wcm/connect/a0f55458-988a-4756-8ebd-f456235bc644/IFC_CTF_Mexico.pdf?MOD=AJPERES&CVID=kCCelk9
3. Campbell, Howard, Leigh Binford, Miguel Bartolomé, and Alicia Barabas. 1993. *Zapotec Struggle: Histories, Politics, and Representations from Juchitán, Oacaca*. Washington: Smithsonian Institution Press.
4. Dunlap, Alexander. 2019. *Renewing Destruction: Wind Energy Development, Conflict and Resistance in a Latin American Context*. London: Rowman & Littlefield.
5. Ibid.
6. Manzo, Carlos. 2011. *Comunalidad, resistencia indígena y neocolonialismo en el Istmo de Tehuantepec, iglos XVI-XXI*. Mexico City: Ce-Acatl; Call, Wendy. 2011. *No Word for Welcome: The Mexican Village Faces the Global Economy*. Lincoln: University of Nebraska Press.
7. UNEP. 2015. "Mexico's Pathways to a Green Economy." United Nations Environment Programme. Accessed 19 November 2015. http://staging.unep.org/greeneconomy/AdvisoryServices/CountryProfiles/Mexico/tabid/104141/Default.aspx
8. Dunlap, Alexander, and Martín Correa-Arce. 2022. "'Murderous Energy' in Oaxaca, Mexico: Wind Factories, Territorial Struggle and Social Warfare." *Journal of Peasant Studies* 49 (2): 455–480.
9. Interview, La Ventosa, 55, 2015.
10. Dunlap, *Renewing Destruction*.
11. Torres Contreras, G.A. 2022. "Who Owns the Land Owns the Wind? Land and Citizenship in the Isthmus of Tehuantepec, Mexico." *Journal of Agrarian Change*: 4. Note that stats used by RAN (National Agricultural Registry) is the same statistic being used ten years ago (cf. Dunlap, ibid.). This statistic has likely changed and been reduced.

12. Dunlap and Correa-Arce, "'Murderous Energy' in Oaxaca, Mexico."
13. Dunlap, *Renewing Destruction*, p. 99; Interview, Juchitán, 1, 2015.
14. Torres Contreras, "Who Owns the Land Owns the Wind?," p. 10.
15. Kass, H. 2022. "Food Anarchy and the State Monopoly on Hunger." *The Journal of Peasant Studies*: 1–20.
16. Interview, La Ventosa, 53, 2015.
17. Interview, La Ventosa, 55, 2015.
18. Dunlap, *Renewing Destruction*.
19. Interview, La Ventosa, 1, 2015.
20. See Dunlap, *Renewing Destruction* or Dunlap, Alexander. 2017. "'The Town Is Surrounded': From Climate Concerns to Life under Wind Turbines in La Ventosa, Mexico." *Human Geography* 10 (2): 16–36.
21. Dunlap, *Renewing Destruction*, p. 54.
22. Interview, La Ventosa 54, 2015.
23. Interview, La Ventosa, 32, 2015.
24. Interview, La Ventosa, 55, 2015.
25. See Dunlap, *Renewing Destruction*.
26. Torres Contreras, G.A. 2021. "Twenty-Five Years under the Wind Turbines in La Venta, Mexico: Social Difference, Land Control and Agrarian Change." *The Journal of Peasant Studies*: 1–19.
27. Smil, Vaclav. 2016. *Energy Transitions: Global and National Perspectives*. Santa Barbara: Praeger.
28. Smith, Patrick. "Soaring Copper Prices Drive Wind Farm Crime." *Wind Power Monthly*, www.windpowermonthly.com/article/1281864/soaring-copper-prices-drive-wind-farm-crime
29. Smil, Vaclav. 2016. *Energy Transitions: Global and National Perspectives*. Santa Barbara: Praeger.
30. Dunlap, Alexander. 2021. "Does Renewable Energy Exist? Fossil Fuel+ Technologies and the Search for Renewable Energy." In: Susana Batel and David Philipp Rudolph(eds), *A Critical Approach to the Social Acceptance of Renewable Energy Infrastructures—Going Beyond Green Growth and Sustainability*. London: Palgrave, pp. 83–102; Tabassum-Abbasi, M. Premalatha, Tasneem Abbasi, and S.A. Abbasi. 2014. "Wind Energy: Increasing Deployment, Rising Environmental Concerns." *Renewable and Sustainable Energy Reviews* 31 (1): 270–288.
31. Policía Auxiliar, Bancaria, Industrial y Comercial (PABIC).
32. Interview, Bíi Hioxo, 2, 2015.
33. Interview, Juchitán, 5, 2015.
34. Ledec, George C., Kennan W. Rapp, and Roberto G. Aiello. 2011. "Greening the Wind: Environmental and Social Considerations for Wind Power Development in Latin America and Beyond." *Energy Unit Sustainable Development Department Latin America and the Caribbean Region The World Bank and Energy System Management Assistance Program (ESMAP)*. www-wds.worldbank.org/external/default/WDSContentServer/WDSP/IB/

2011/07/26/000333038_20110726003613/Rendered/PDF/634800v10WP0GrooBOX361518BooPUBLICo.pdf

35. Dunlap, *Renewing Destruction.*
36. Abbasi, SA, Tabassum-Abbasi, and Tasneem-Abbasi. 2016. "Impact of Wind-Energy Generation on Climate: A Rising Spectre." *Renewable and Sustainable Energy Reviews* 59: 1591–1598.
37. Lucio, Carlos Federico. 2016. *Conflictos Socioambientales, Derechos Humanos Y Movimiento Indígena En El Istmo De Tehuantepec.* Zacatecas: Universidad Autónoma de Zacatecas.
38. Union Hidalgo, 4, 2019.
39. Dunlap, *Renewing Destruction*; Dunlap and Correa-Arce, "'Murderous Energy' in Oaxaca, Mexico."
40. See Dunlap, *Renewing Destruction*, for a greater conversation on health findings.
41. See Dunlap, Alexander. 2017. "'The Town Is Surrounded': From Climate Concerns to Life under Wind Turbines in La Ventosa, Mexico." *Human Geography* 10 (2): 16–36.
42. The Comisión Federal de Electricidad.
43. Dunlap, "The Town is Surrounded."
44. *Anti-politics* recognizes politics in-and-of itself as a method of social control, stupidification and obstruction to social change. This entails the political position that rejects political mediation through political parties, participation in parliament or collaboration with governmental authorities in general. The anti-political position, moreover, denies political institutions any role in the post-revolutionary period. The anti-political position, or struggle, is by every means political—in the most general sense of the term—but it rejects politics as a method of social war and control. See the works by Alfredo M. Bonanno, Jean Weir, Gustavo Rodríguez, Wolfi Landstreicher, Sasha K., Mauricio Morales and many more.
45. Dunlap, *Renewing Destruction.*
46. Ibid.
47. Ibid.
48. Dunlap, Alexander. 2022. "Conclusion: A Call to Action, Towards an Insurrection in Energy Research." In: Majia H. Nadesan, Martin J. Pasqualetti, and Jennifer Keahey (eds), *Energy Democracies for Sustainable Futures.* Amsterdam: Academic Press.
49. See Marya, Rupa, and Raj Patel. 2021. *Inflamed: Deep Medicine and the Anatomy of Injustice.* New York: Picador.
50. Ibid.
51. Interview, 13, 1 January 2020.
52. Dunlap, *Renewing Destruction.*
53. Dunlap and Correa-Arce, "'Murderous Energy' in Oaxaca, Mexico," p. 459.

54. Dunlap, Alexander. 2018. “Insurrection for Land, Sea and Dignity: Resistance and Autonomy against Wind Energy in Álvaro Obregón, Mexico.” *Journal of Political Ecology* 25: 130.
55. Ibid., p. 131.
56. See endnote 10, in Dunlap, Alexander. 2020. “Wind, Coal, and Copper: The Politics of Land Grabbing, Counterinsurgency, and the Social Engineering of Extraction.” *Globalizations* 17 (4): 661–682.
57. Dunlap & Correa-Arce, “‘Murderous Energy’ in Oaxaca, Mexico,” p. 471.
58. Ibid.
59. Dunlap, Alexander. 2019. “Revisiting the Wind Energy Conflict in Gųi’xhi’ Ro / Álvaro Obregón: Interview with an Indigenous Anarchist.” *Journal of Political Ecology* 26 (1): 165.
60. See Torres Contreras, “Twenty-Five Years under the Wind Turbines in La Venta, Mexico and “Who Owns the Land Owns the Wind?”
61. See Hofmann, Susanne. “The Interoceanic Corridor of the Isthmus of Tehuantepec Infrastructure Project.”

CHAPTER 3: FIGHTING THE WORLDEATER

1. Interview, 15 August 2022.
2. Brock, A. and A. Dunlap 2018. “Normalising Corporate Counterinsurgency: Engineering Consent, Managing Resistance And Greening Destruction Around the Hambach Coal Mine and Beyond.” *Political Geography* 62(1), pp. 33–47.
3. Malone, Robert. 12 March 2007. “The World’s Biggest Land Vehicle.” *Forbes Magazine*. Available at: www.forbes.com/2007/03/12/bagger-vehicle-tractor-biz-logistics-cx_rm_0312vehicle.html
4. Wikipedia. (n.d.) *Bagger 288*. Wikipedia. Available at: https://en.wikipedia.org/wiki/Bagger_288#cite_note-2
5. Ibid.
6. Brock and Dunlap, “Normalising Corporate Counterinsurgency,” p. 34.
7. Ibid.
8. Ibid. And also see: Brock, A. 2020. “Securing accumulation By Restoration—Exploring Spectacular Corporate Conservation, Coal Mining and Biodiversity Compensation in the German Rhineland.” *Environment and Planning E: Nature and Space*: 2514848620924597.
9. Interview, 15 August 2022.
10. Geronimo. 2012. *Fire and Flames: A History of the German Autonomist Movement*. Oakland: PM Press.
11. Interview, 15 August 2022.
12. For more on the anti-globalization movement and “summit hopping” see Graeber, David. 2009. *Direct Action: An Ethnography*. Oakland: AK press.
13. Black Bloc tactics emerged to challenge the politics and armed vanguardism of the Red Army Faction (RAF) and other Marxist-Leninist groups, to

instead allow the mass participation of militant action in demonstrations by wearing masks and directly attacking political targets. The idea was to create popular and anti-authoritarian militancy rooted in neighborhoods and local struggles, to promote autonomy instead of Communist party politics and command and control structures. For more see: Dupuis-Déri, Francis. 2014. *Who's Afraid of the Black Blocs?: Anarchy in Action around the World*. Oakland: PM Press; Chapter 11, "Meanwhile, Elsewhere on the Left ..." In: Smith, J, and André Moncourt. 2009. *The Red Army Faction, a Documentary History: Volume 1: Projectiles for the People*. Oakland: PM Press.

14. Leila, and Scacha. 2001. "The Anarchist Ethic in the Age of the Anti-Globalization Movement." *Killing King Abacus* 2: 1–20. https://archive.elephanteditions.net/library/killing-king-abacus-the-anarchist-ethic-in-the-age-of-the-anti-globalisation-movement.pdf and Graeber, 2009.
15. Müller, Tadzio. 2022. "Climate Justice: Global Resistance to Fossil-Fueled Capitalism." In Nathan Barlow, Liva Regen, Noemie Cadiou, Ekaterina Chertkovskaya, Max Hollweg, Christina Plank, Merle Schulken, and Verena Wolf (eds), *Degrowth and Strategy: How to Bring About Social-Ecological Transformation*. London: Mayfly, pp. 114–127.
16. Zee, Bibi van der. "Activists Reveal Tactics Used by Police to 'Decapitate' Copenhagen Climate Protests." *The Guardian*. www.theguardian.com/environment/2009/dec/17/copenhagen-police-tactics-revealed
17. I altered the quote to read "COP15" instead of COP17 as they wrote in the original, as this author seems to be miss-remembering the COP events. I rely on this list and number of COP events, here: www.downtoearth.org.in/climate-change/coplist
18. Müller, "Climate Justice," p. 121.
19. X, Andrew. "Give up Activism. Do or Die." https://theanarchistlibrary.org/library/andrew-x-give-up-activism.
20. Ibid.
21. Ibid., p. 7.
22. Anonymous. "Hambach Forest Occupation," *Machorka* p. 15. https://machorka.espivblogs.net/2015/07/13/new-zine-hambach-forest-occupation
23. Ibid., p. 4.
24. Ibid.
25. Brock and Dunlap, "Normalising Corporate Counterinsurgency."
26. Anonymous, "Hambach Forest Occupation," p. 16.
27. Ibid.
28. Wolfe, Patrick. 1999. *Settler Colonialism and the Transformation of Anthropology*. London: Routledge; see also: Dunlap, Alexander. 2018. "The 'Solution' Is Now the 'Problem': Wind Energy, Colonization and the 'Genocide-Ecocide Nexus' in the Isthmus of Tehuantepec, Oaxaca." *The International Journal of Human Rights* 42 (4): 550–573.
29. Churchill, Ward. 2012. "Confronting Western Colonialism, American Racism, and White Supremacy: Ward Churchill and Pierre Orelus in

Dialogue." In Pierre W. Orelus (ed), *A Decolonizing Encounter: Ward Churchill and Antonia Darder in Dialogue*. New York Peter Lang, pp. 56–112.

30. Foucault, Michel. 2003 (1997).*"Society Must Be Defended:" Lectures at the College De France 1975–1976*. New York: Picador.
31. Arendt, Hannah. 1962 [1951]. *The Origins of Totalitarianism*. New York: The World Publishing Company, 1962 [1951]; Césaire, Aimé. 2001 [1955]. *Discourse on Colonialism*. New York: NYU Press.
32. Brock and Dunlap, "Normalising Corporate Counterinsurgency."
33. Ibid., p. 37.
34. Ibid.
35. Neubauer, G. 2013. *Schwarzbuch Kohlepolitik*. Hamburg: Greenpeace.
36. Brock and Dunlap, "Normalising Corporate Counterinsurgency," p. 37.
37. Ibid.
38. Ibid.
39. Ibid.
40. Brock, Andrea. 2023. "Life at Lutzerath." *The Ecologist*. https://theecologist.org/2023/jan/19/life-lutzerath
41. Brock and Dunlap, "Normalising Corporate Counterinsurgency," p. 39.
42. Ibid.
43. Ibid.
44. Wong, Julia Carrie. 2016. "Dakota Access Pipeline: 300 Protesters Injured after Police Use Water Cannons." *The Guardian*, www.theguardian.com/us-news/2016/nov/21/dakota-access-pipeline-water-cannon-police-standing-rock-protest See more on the NoDAPL struggle see: Estes, Nick. 2019. *Our History Is the Future: Standing Rock Versus the Dakota Access Pipeline, and the Long Tradition of Indigenous Resistance*. New York: Verso; Brown, Alleen, Will Parrish, and Alice Speri. 2017. "Leaked Documents Reveal Counterterrorism Tactics Used at Standing Rock to 'Defeat Pipeline Insurgencies.' [Article Series Contains Four Parts]." *The Intercept*. https://theintercept.com/2017/05/27/leaked-documents-reveal-security-firms-counterterrorism-tactics-at-standing-rock-to-defeat-pipeline-insurgencies/ and Granovsky-Larsen, Simon, and Larissa Santos. 2021. "From the War on Terror to a War on Territory: Corporate Counterinsurgency at the Escobal Mine and the Dakota Access Pipeline." *Canadian Journal of Latin American and Caribbean Studies/Revue canadienne des études latino-américaines et caraïbes*: 1–25.
45. RWE. "Recultivating Open-Cast Mines." *RWE*. www.rwe.com/en/responsibility-and-sustainability/environmental-protection/recultivating-open-cast-mines
46. RWE. "Climate Protection." *RWE*. www.rwe.com/en/responsibility-and-sustainability/environmental-protection/climate-protection
47. Brock, Andrea. 2018. *Conserving Power: An Exploration of Biodiversity Offsetting in Europe and Beyond*. Brighton: University of Sussex: Doctoral Thesis.

48. Brock, Andrea. 2020. "Securing Accumulation by Restoration—Exploring Spectacular Corporate Conservation, Coal Mining and Biodiversity Compensation in the German Rhineland." *Environment and Planning E: Nature and Space*: 1–32.
49. Brock and Dunlap, "Normalising Corporate Counterinsurgency."
50. Dunlap, Alexander. 2021. "Does Renewable Energy Exist? Fossil Fuel+ Technologies and the Search for Renewable Energy." In *A Critical Approach to the Social Acceptance of Renewable Energy Infrastructures—Going Beyond Green Growth and Sustainability*, edited by Susana Batel and David Philipp Rudolph, 83–102. London: Palgrave.
51. Brock and Dunlap, "Normalising Corporate Counterinsurgency."
52. Ibid., p. 40.
53. For more on offsetting read: Sullivan, Sian. 2009. "Green Capitalism, and the Cultural Poverty of Constructing Nature as Service Provider." *Radical Anthropology* 3 (1): 18–27; Sullivan, Sian. 2017. "What's Ontology Got to Do with It? On Nature and Knowledge in a Political Ecology of The 'Green Economy.'" *Journal of Political Ecology* 24: 217–242.
54. Brock and Dunlap, "Normalising Corporate Counterinsurgency," p. 40.
55. Brock, "Securing accumulation By Restoration," p. 10.
56. Huff, Amber, and Andrea Brock. 2017. "Intervention—'Accumulation by Restoration: Degradation Neutrality and the Faustian Bargain of Conservation Finance.'" *Antipode Foundation*, https://antipodefoundation.org/2017/11/06/accumulation-by-restoration/
57. Brock, "Securing accumulation By Restoration," p. 9.
58. Ibid., p. 18.
59. Peter, Claudia. 2002. "Digging up Astroturf." In *Battling Big Business: Countering Greenwash, Infiltration, and Other Forms of Corporate Bullying*, edited by Eveline Lubbers, 147–154. Devon: Green Books.
60. Brock and Dunlap, "Normalising Corporate Counterinsurgency," p. 39.
61. Mohr, Almut, and Mattijs Smits. 2022. "Sense of Place in Transitions: How the Hambach Forest Movement Shaped the German Coal Phase-Out." *Energy Research & Social Science* 87: 1–12; Liersch, Carina, and Peter Stegmaier. 2022. "Keeping the Forest above to Phase out the Coal Below: The Discursive Politics and Contested Meaning of the Hambach Forest." *Energy research & social science* 89: 1–11; Kalt, Tobias. 2021. "Jobs Vs. Climate Justice? Contentious Narratives of Labor and Climate Movements in the Coal Transition in Germany." *Environmental Politics* 30 (7): 1135–1154; Henseleit, Meike, Sandra Venghaus, and Wilhelm Kuckshinrichs. 2022. "The Hambach Forest in the German Debate on Climate Protection: Is There a Symbolic Value Beyond the Environmental Value?" In *Sustainable Rural Development*: IntechOpen. This article offers a mild exception, recognizing in passing sabotage and the influence of the Earth Liberation Front (ELF) and eco-anarchist actions. Kaufer, Ricardo, and Paula Lein. 2020. "Anarchist

Resistance in the German Hambach Forest: Localising Climate Justice." *Anarchist studies* 28 (1): 60–83.
62. Treu, Nina, Matthias Schmelzer, and Corinna Burkhart. 2020. *Degrowth in Movement (S): Exploring Pathways for Transformation*. New York: John Hunt Publishing.
63. Churchill, Ward. 2007 [1998]. *Pacifism as Pathology: Reflections on the Role of Armed Struggle in North America*. Oakland: AK Press.
64. Gelderloos, Peter. 2013. *How Nonviolence Protects the State*. Cambridge: South End Press, 2007; Gelderloos, Peter. *The Failure of Nonviolence: From Arab Spring to Occupy*. Seattle: Left Bank Books.
65. Osterweil, Vicky. 2019. *In Defense of Looting: A Riotous History of Uncivil Action*. New York: Bold Type Books.
66. McBay, Aric. 2019. *Full Spectrum Resistance, Vol. 1 and 2: Building Movements and Fighting to Win*. New York: Seven Stories Press.
67. Gelderloos, *How Nonviolence Protects the State*.
68. Ibid.
69. Brock, "Life at Lutzerath."
70. Ibid.
71. Anonymous. 2016. "Text Concerning Hambach Forest (Germany)." *Return Fire* 3: 91–92. https://returnfire.noblogs.org/files/2021/03/Return-Fire-vol.3-chap.3-pg68-end.pdf
72. From Wikipedia: "Tree spiking involves hammering a metal rod, nail or other material into a tree trunk, either inserting it at the base of the trunk where a logger might be expected to cut into the tree, or higher up where it would affect the sawmill later processing the wood."
73. Anonymous, "Text Concerning Hambach Forest (Germany)," p. 91.
74. Anonymous. "Nuclear or Lignite: Fragments For the Struggle against This World's Juice—from Teh Bois Lejuc to the Hambach Forest..." *Act for Freedom*. https://lille.indymedia.org/IMG/pdf/nuclear_or_lignite_fragments_of_the_struggle_against_this_world_s_juice_from_the_bois_lejuc_to_the_hambach_forest.page_by_page.pdf
75. Ibid.
76. Ibid.
77. Kaufer and Lein, "Anarchist Resistance in the German Hambach Forest: Localising Climate Justice."
78. Wecker, Katharina. "Hambach in Shock after Journalist's Death." *DW*. www.dw.com/en/stillness-and-shock-in-hambach-forest-after-journalist-dies/a-45579629
79. Brock, "Life at Lutzerath."

CHAPTER 4: MINERAL DEMAND

1. Hancco, Nelly. "Arequipa: Un 'Trozo' De Marte Está En El Desierto De La Joya." *Correo*. https://diariocorreo.pe/edicion/arequipa/arequipa-un-trozo-de-marte-esta-en-el-desierto-de-la-joya-682732/

2. Jaskoski, Maiah. 2014. "Environmental Licensing and Conflict in Peru's Mining Sector: A Path-Dependent Analysis." *World Development* 64: 873–883; Romero, José Antonio Lapa. 2017. *Lo Que Los Ojos No Ven: Capital Minero, Hegemonia, Represión Estatal Y Movimiento Social En El Valle De Tambo De Marzo a Mayo Del 2015: El Caso Del Conflicto Tia Maria En La Región Arequipa*. Lima: Grupo Editorial Arteidea.
3. Stern, Steve J. 1998. *Shining and Other Paths: War and Society in Peru, 1980–1995*. Durham, NC: Duke University Press.
4. Ibid.
5. Ibid.
6. Fumerton, Mario A. 2018. "Beyond Counterinsurgency: Peasant Militias and Wartime Social Order in Peru's Civil War." *European Review of Latin American and Caribbean Studies/Revista Europea de Estudios Latinoamericanos y del Caribe* 105: 61–86.
7. Anonymous. 2023. "The Uprising in Peru: Anarchists Discuss the Revolt against Police Violence and the State of Emergency." *Crimethinc*. https://crimethinc.com/2023/02/19/the-uprising-in-peru-popular-revolt-against-police-violence-and-the-state-of-emergency
8. Villalba, Fernando Velásquez. 2022. "A Totalidade Neoliberal-Fujimorista: Estigmatização E Colonialidade No Peru Contemporaneo" *Revista Brasileira de Ciências Sociais* 37: 1–16.
9. Ibid.
10. Dunlap, Alexander. 2014. "Permanent War: Grids, Boomerangs, and Counterinsurgency." *Anarchist Studies* 22 (2): 55–79.
11. Anonymous, "The Uprising in Peru."
12. Dunlap, Alexander, and Hannah Kass. "Stop Cop City! Revisiting Degrowth & Permanent Ecological Conflict." *Degrowth Info*, https://degrowth.info/en/blog/stop-cop-city-revisiting-degrowth-permanent-ecological-conflict
13. Pratt, Timothy. 2023. "'Assassinated in Cold Blood': Activist Killed Protesting Georgia's 'Cop City.'" *The Guardian*. www.theguardian.com/us-news/2023/jan/21/protester-killed-georgia-cop-city-police-shooting
14. McClintock, Cynthia. 2005. "The Evolution of Internal War in Peru: The Conjunction of Need, Creed, and or Gan I Za Tion Al Finance." In: Cynthia J. Arnson and I. William Zartman (eds), *Rethinking the Economics of War: The Intersection of Need, Creed, and Greed*. Baltimore: Johns Hopkins University Press, pp. 52–83.
15. Jaskoski, Maiah. 2013. *Military Politics and Democracy in the Andes*. Baltimore: Johns Hopkins University Press, p. 65.
16. Ibid.
17. Fumerton, "Beyond Counterinsurgency."
18. Jaskoski, *Military Politics and Democracy in the Andes*, p. 75.
19. Ibid.
20. Ibid., p. 168.
21. Ibid., p. 171.

22. Middeldorp, Nick, Carlos Morales, and Gemma van der Haar. 2016. "Social Mobilisation and Violence at the Mining Frontier: The Case of Honduras." *The Extractive Industries and Society* 3 (4): 934.
23. Andreucci, Diego, and Giorgos Kallis. 2017. "Governmentality, Development and the Violence of Natural Resource Extraction in Peru." *Ecological Economics* 134 (C): 95–103.
24. Grufides. "Police in the Pay of Mining Companies: The Responsibility of Switzerland and Peru for Human Rights Violations in Mining Disputes." *Grufides*, p. 9. www.grufides.org/sites/default/files//documentos/documentos/Report%2520Eng.compressed.pdf
25. Ibid.
26. Ibid., p. 10.
27. Ibid.
28. Jaskoski, *Military Politics and Democracy in the Andes.*
29. Gustafson, Bret, and Natalia Guzmán Solano. 2016. "Mining Movements and Political Horizons in the Andes: Articulation, Democratisation, and Worlds Otherwise." In: Kalowatie Deonandan and Michael L. Dougherty (eds), *Mining in Latin America: Critical Approaches to the New Extraction*. London: Routledge, pp. 141–159.
30. Bravo, Romina Rivera. 2015. "Militarization and Environmental Struggles in Peru." *SOA Watch*, p. 1. www.soaw.org/about-us/equipo-sur/200-encuentro/4287-perufeb2015
31. Paley, Dawn. 2014. *Drug War Capitalism*. Oakland: AK Press.
32. FLD. 2017. "UPR Submission—Peru 2017." *Frontline Defenders*. www.frontlinedefenders.org/en/statement-report/upr-submission-peru-2017
33. Palomino, Roy. 2015. "[Peru] Estas Son Las Mineras Quele Pagan a Policía Para Que Repriman Las Protestas Contra Ellas." *Mi Mina Corrupta*. https://minacorrupta.wordpress.com/tag/dinoes/
34. Lust, Jan. 2014. "Peru: Mining Capital and Social Resistance." In: Henry Veltmeyer and James F Petras (eds), *The New Extractivism: A Post-Neoliberal Development Model or Imperialism of the Twenty-First Century?* London: Zed Books, 192–221.
35. MEM. "Inversión Minera Se Acelera Y Crece En 54.8% En Noviembre." *Minerstry of Energy and Mines*, www.minem.gob.pe/_detallenoticia.php?idSector=1&idTitular=8244
36. See Bebbington, Anthony. 2012. *Social Conflict, Economic Development and the Extractive Industry: Evidence from South America*. London: Routledge; Bebbington, Anthony, Denise Humphreys Bebbington, Jeffrey Bury, Jeannet Lingan, Juan Pablo Muñoz, and Martin Scurrah. 2008. "Mining and Social Movements: Struggles over Livelihood and Rural Territorial Development in the Andes." *World Development* 36 (12): 2888–2905; Bury, Jeffrey, and Timothy Norris. 2014. "Rocks, Rangers, and Resistance: Mining and Conservation Frontiers in the Cordillera Huayhuash, Peru." In: Anthony Bebbington and Jeffrey Bury (eds), *Subterranean Struggles: New Dynamics*

of Mining, Oil, and Gas in Latin America. Austin: University of Texas press; Arce, Moisés. *Resource Extraction and Protest in Peru*. Pittsburgh: University of Pittsburgh Press.

37. Fernández, Marlene Castillo, Mary Chávez Quijada, Mirella Gallardo Marticorena, Equipo Técnico de la Red Muqui, David del Carpio Lazo, and Jesús Gómez Urquizo. 2011. "Valle De Tambo-Islay. Territorio, Agua Y Derechos Locales En Riesgo Con La Minería a Tajo Abierto." *Copper Accion*. http://cooperaccion.org.pe/publicaciones/valle-de-tambo-islay-territorio-agua-y-derechos-locales-en-riesgo-con-la-mineria-a-tajo-abierto/
38. Lust, "Peru: Mining Capital and Social Resistance."
39. Kirsch, Stuart. 2014. *Mining Capitalism: The Relationship between Corporations and Their Critics*. Berkeley: University of California Press.
40. Southern Copper Peru. www.youtube.com/watch?v=Pd1OL9EEj4k
41. See minutes 6:30–7:00 of RTVE (2016). *La Battalla del Cobre*. www.rtve.es/television/20160823/batalla-del-cobre/1296701.shtml
42. Interviews 1, 30, and 37.
43. Frente Amplio de defense y desarrollo de los intereses de la provincial de Islay.
44. Bebbington, *Social Conflict, Economic Development and the Extractive Industry*, p. 92.
45. http://larepublica.pe/archivo/704641-tacna-fiscalia-investiga-a-southern-por-presunta-contaminacion. http://larepublica.pe/politica/978337-pasto-grande-pide-demolicion-de-cancha-de-relaves-de-southern. www.radiouno.pe/noticias/23662/viceministros-se-negaron-ingresar-embalse-relaves-quebrada-honda
46. See https://ejatlas.org/conflict/cuajone-toquepala-ilo-peru
47. http://elpueblo.com.pe/noticia/opinion/quien-mato-carlos-guillen-carrera and www.youtube.com/watch?v=TvC5PARFWGg
48. Romero, *Lo Que Los Ojos No Ven*.
49. Ibid.
50. Ibid., p. 21.
51. Ibid., pp. 21–30.
52. Interview 11, 13 January 2018; Interview 20, 15 January 2018; Interview 34, 17 January 2018.
53. Nixon, Rob. 2011. *Slow Violence and the Environmentalism of the Poor*. Cambridge: Harvard University Press; Springer, Simon and Philippe Le Billon. 2016. "Violence and Space: An Introduction to the Geographies of Violence." *Political Geography* 52: 1–3; Davies, Thom. 2018. "Toxic Space and Time: Slow Violence, Necropolitics, and Petrochemical Pollution." *Annals of the American Association of Geographers* 108 (6): 1537–1553.
54. Romero, *Lo Que Los Ojos No Ven*, p. 46.
55. Ibid.
56. Bebbington, *Social Conflict, Economic Development and the Extractive Industry*; Dunlap, 2019; Gamu, Jonathan Kishen, and Peter Dauvergne. 2018.

"The Slow Violence of Corporate Social Responsibility: The Case of Mining in Peru." *Third World Quarterly* 39 (5): 959–975.

57. Interviews and Jaskoski, "Environmental Licensing and Conflict in Peru's Mining Sector."
58. Arce, *Resource Extraction and Protest in Peru.*
59. Sullivan, Lynda. 2015. "Peru's Tia Maria Mining Conflict: Another Mega Imposition." *Upside Down World*, http://upsidedownworld.org/archives/peru-archives/perus-tia-maria-mining-conflict-another-mega-imposition/
60. Castillo Fernández et al., "Valle De Tambo-Islay"; Jaskoski, "Environmental Licensing and Conflict in Peru's Mining Sector."
61. Sullivan, "Peru's Tia Maria Mining Conflict."
62. Castillo Fernández et al., "Valle De Tambo-Islay", pp. 80–83; Sullivan, "Peru's Tia Maria Mining Conflict."
63. Sullivan, "Peru's Tia Maria Mining Conflict."
64. Interview 24, 16 January 2018.
65. Ibid.
66. Romero, *Lo Que Los Ojos No Ven.*
67. Interview 47, 18 February 2018.
68. Romero, *Lo Que Los Ojos No Ven*; Sullivan, "Peru's Tia Maria Mining Conflict."
69. Daly, Gabriel. "Tía María: Los Factores Detrás Del Conflicto (Informe)." *El Comercio*. https://elcomercio.pe/peru/arequipa/tia-maria-factores-detras-conflicto-informe-351505
70. Aguirre, Carlos. 2011. "Terruco De M … Insulto Y Estigma En La Guerra Sucia Peruana." *Histórica* 35 (1): 103–139.
71. Bebbington, *Social Conflict, Economic Development and the Extractive Industry*; Kallis and Andreucci, 2017; Anonymous, "The Uprising in Peru."
72. Wiener, Gabriela. "Apología De La Imbecilidad." *La República*.https://larepublica.pe/politica/1175676-apologia-a-la-imbecilidad
73. Trelles, Nelly. "Terruquear, Terruqueo, Terruqueadores." *Castellano Actual, Universidad de Piura*. http://udep.edu.pe/castellanoactual/terruquear-terruqueo-terruqueadores/
74. La Republica. "Tía María: Southern pide disculpas a población del Valle de Tambo por insinuar que son terroristas." https://larepublica.pe/sociedad/1247051-tia-maria-southern-pide-disculpas-poblacion-valle-tambo-insinuar-son-terroristas
75. Romero, *Lo Que Los Ojos No Ven.*
76. Interview 10, 13 January 2018.
77. See Mollendinostv. 2016. "Ollanta Humala ¿traicionó a su palabra?" *Consulta*. www.youtube.com/watch?v=0Fr2PEVToWY; Reuters. 2018. *La Battalla del Cobre*. www.rtve.es/television/20160823/batalla-del-cobre/1296701.shtml
78. ee Tv Perú. 2015. "Mensaje a la Nación del Presidente Ollanta Humala." www.youtube.com/watch?v=bselvRy_cjY
79. Romero, *Lo Que Los Ojos No Ven*, p. 54.

80. See La República. 2017. http://larepublica.pe/politica/1154229-gra-recibira-donacion-de-s770-mil-de-southern
81. See BBC (2017) Odebrecht case: Politicians worldwide suspected in bribery scandal. Available at: http://www.bbc.com/news/world-latin-america-41109132
82. Interview 26, 16 January 2018.
83. Interview 24, 16 January 2018.
84. Interview 43, 18 January 2018.
85. La República. 2015. "Video muestra que policía 'sembró' arma a manifestante contra Tía María | VIDEO." http://larepublica.pe/politica/872285-video-muestra-que-policia-sembro-arma-a-manifestante-contra-tia-maria-video
86. Redacción. 2017. "Absuelven a agricultor a quien 'sembraron' arma durante protestas contra Tía María.": http://rpp.pe/peru/arequipa/absuelven-a-agricultor-a-quien-sembraron-arma-durante-protestas-contra-tia-maria-noticia-1071324
87. Interview 42, 18 January 2018.
88. Interview 2.2, 19 January 2018.
89. Interview 12, 14 January 2018.
90. Interview 17, 15 January 2018.
91. Interview 40, 18 January 2018.
92. Jaskoski, *Military Politics and Democracy in the Andes.*
93. Arendt, Hannah. 1962 [1951]. *The Origins of Totalitarianism.* New York: The World Publishing Company, p. 289.
94. Interview 1, 2, 17, 19, 26, 40, and 41.
95. Interview 19, 15 January 2018.
96. Palomino, "[Peru] Estas Son Las Mineras Quele Pagan a Policía Para Que Repriman Las Protestas Contra Ellas."
97. FN, 14 January 2018; Interview 2.2, 12 January 2018.
98. Interview 12, 14 January 2018.
99. Wilson, Japhy. 2023. *Extractivism and Universality: Inside an Uprising in the Amazon.* Oxon: Routledge.
100. Interview 2, 12 January 2018.
101. La Republica. 2018. "Tía María: Piden 30 años de prisión para Pepe Julio Gutiérrez." http://larepublica.pe/sociedad/1172269-piden-30-anos-de-prision-para-pepe-julio-gutierrez
102. Romero, *Lo Que Los Ojos No Ven*, p. 99.
103. Kitson, Frank. 2010 [1971]. *Low Intensity Operations: Subversion, Insurgency, and Peace Keeping.* London: Bloomsbury House, pp. 69–71. On "neutralization," see also: Churchill, Ward. 2002 [1988]. *Agents of Repression: The Fbi's Secret Wars against the Black Panther Party and the American Indian Movement.* Cambridge, MA: SouthEnd Press, p. 44.
104. Interview 36, 17 January 2018.
105. This quote is fragmented to prevent the identification of this research participant.

106. Interview 36, 17 January 2018.
107. Interview 1, 12 January 2018.
108. Dunlap, Alexander, and James Fairhead. 2014. "The Militarisation and Marketisation of Nature: An Alternative Lens to 'Climate-Conflict.'" *Geopolitics* 19 (4): 937–961; Verweijen, Judith, and Esther Marijnen. 2018. "The Counterinsurgency/Conservation Nexus: Guerrilla Livelihoods and the Dynamics of Conflict and Violence in the Virunga National Park, Democratic Republic of the Congo." *The Journal of Peasant Studies* 45 (2): 300–320; Brock, Andrea. 2020. "'Frack Off': Towards an Anarchist Political Ecology Critique of Corporate and State Responses to Anti-Fracking Resistance in the UK." *Political Geography* 82: 102246; Verweijen, Judith, and Alexander Dunlap. 2021. "The Evolving Techniques of Social Engineering, Land Control and Managing Protest against Extractivism: Introducing Political (Re)Actions 'from Above.'" *Political Geography* 83: 1–9.
109. FM3-24. "Insurgencies and Countering Insurgencies." *Department of the Army Headquarters*, http://fas.org/irp/doddir/army/fm3-24.pdf
110. Ibid., p. 10–11. [It is a military manual so the pages are different].
111. Koven, Barnett S. 2016. "Emulating Us Counterinsurgency Doctrine: Barriers for Developing Country Forces, Evidence from Peru." *Journal of Strategic Studies* 39 (5–6): 878–898.
112. FM 3-24, p. 8-1.
113. Interview 24, 16 January 2018.
114. Romero, *Lo Que Los Ojos No Ven.*
115. Williams, Kristian. 2007. *Our Enemies in Blue: Police and Power in America.* Cambridge: South End Press, pp. 211–212.
116. Ibid., p. 58.
117. Ibid.
118. On corporate culture, please read: Dugger, William. 1989. *Corporate Hegemony.* Connecticut: Greenwood Press.
119. Interview 24, 16 January 2018.
120. Interview 24, 16 January 2018.
121. Southern. 2016. *Construyamos Confianza: Proyecto Tía Maria.* Peru: Southern Copper Peru, pp. 1–55.
122. Price, Dave H. 2014. "Counterinsurgency by Other Names: Complicating Humanitarian Applied Anthropology in Current, Former, and Future War Zones." *Human Organization* 73 (2): 95–105.
123. Wilson, *Extractivism and Universality*, p. 55.
124. Bernays, Edward. 1947. "The Engineering of Consent." *The ANNALS of the American Academy of Political and Social Science* 250 (1): 118.
125. Interview 24, 16 January 2018.
126. Hochmüller, Markus, and Markus-Michael Müller. 2017. "Countering Criminal Insurgencies: Fighting Gangs and Building Resilient Communities in Post-War" In: Louise Wiuf Moe and Markus-Michael Müller (eds),

Reconfiguring Intervention: Complexity, Resilience and The "local Turn" in Counterinsurgent Warfare. London: Palgrave MacMillan, p. 175.

127. Interview 24, 16 January 2018.
128. Interviews: 37, 13, and 27 in order of sentence.
129. Interview 28, 16 January 2018.
130. Interview 3, 12 January 2018.
131. Interview 21, 15 January 2018.
132. Polanyi, Karl. 2001 [1947]. *The Great Transformation: The Political and Economic Origins of Our Time*. Boston: Beacon Press, p.118.
133. Bateman, Milford. 2010. *Why Doesn't Microfinance Work? The Destructive Rise of Local Neoliberalism*. New York: Zed Books, p. 74.
134. Interview 43, 18 January 2018.
135. Romero, *Lo Que Los Ojos No Ven*.
136. For discussions on the patterned and generalized use of "Good/Bad" protester as a divisive statist strategy, see: Gelderloos, Peter. 2013. *The Failure of Nonviolence: From Arab Spring to Occupy*. Seattle: Left Bank Books.
137. Interview 10, 13 January 2018.
138. Interview 17, 15 January 2018.
139. Interview 47, 5 February 2018.
140. Anonymous, "The Uprising in Peru"; Wilson, Japhy. 2023. "The Universal Humanity of the Peruvian Uprising." *Undisciplined Environments*. https://undisciplinedenvironments.org/2023/02/07/universal-humanity-peru-uprising/
141. El Popular. 2021. "Dina Boluarte: Tía María No Va, Porque Es Inviable El Tema." *Actualidad El Popular*, https://elpopular.pe/actualidad/2021/09/26/dina-boluarte-tia-maria-va-porque-es-inviable-tema-86017
142. El Búho. 2023. "Tía María: Las Visitas De Southern Perú Al Gabinete Otárola." *El Búho*. https://elbuho.pe/2023/01/arequipa-tia-maria-las-visitas-de-southern-peru-al-gabinete-otarola/
143. RPP. "Dina Boluarte Sobre Tía María: 'Nada Está Cerrado, Pero No Podemos Anteponer El Tema Del Oro Por Encima Del Agua.'" *Redacción RPP*. https://rpp.pe/economia/economia/dina-boluarte-sobre-tia-maria-no-podemos-anteponer-el-tema-del-oro-la-plata-por-encima-del-agua-noticia-1455597?ref=rpp
144. RPP. "Ministro De Energía Y Minas Dice Que El Gobierno Está En 'Diálogo Permanente' Con La Sociedad Civil Para Evitar Protestas." *Redacción RPP*. https://rpp.pe/politica/gobierno/oscar-vera-dice-que-el-gobierno-esta-en-dialogo-permanente-con-la-sociedad-civil-para-evitar-protestas-noticia-1457136?ref=rpp
145. Búho, El. 2022. "Arequipa: Organizador De Perumin 36 Considera Que Rechazo a Tía María Es Un Grave Error." *Redacción El Búho*. https://elbuho.pe/2022/11/arequipa-organizador-de-perumin-36-considera-que-rechazo-a-tia-maria-es-un-grave-error/

146. Búho, El. 2022. "Arequipa: ¿Cuál Es La Posición Del Electo Gobernador Rohel Sánchez Sobre Tía María Y Majes—Siguas Ii?" *Redacción El Búho*. https://elbuho.pe/2022/10/arequipa-cual-es-la-posicion-del-electo-gobernador-rohel-sanchez-sobre-tia-maria-y-majes-siguas-ii/

CHAPTER 5: TRAPPED IN THE GRID

1. Apologies Christophe if I was rude in any way. I greatly appreciate your work and efforts.
2. In French: *Réseau de Transport d'Électricité*.
3. To learn more, I recommend the book: *The Hand-Sculpted House: A Practical and Philosophical Guide to Building a Cob Cottage* by Ianto Evans, Mchael G. Smith, and Linda Smiley.
4. Check out the ENTSO-E. "Entso-E Transmission System Map." *European Network of Transmission System Operators for Electricity*, www.entsoe.eu/data/map/
5. Weber, Eugen. 1976. *Peasants into Frenchmen: The Modernization of Rural France, 1870–1914*. Palo Alto: Stanford University Press.
6. Ibid.; MTC. 2019. "Pas Res Nos Arresta! Préface à l'édition Italienne Du Livre de Gérard De Sède: 700 Ans de Révoltes Occitanes." *Troupe Collectif Mauvaise*. https://mauvaisetroupe.org/spip.php?article217
7. See www.everyculture.com/Europe/Aveyronnais-Economy.html
8. Nadaï, Alain, and Olivier Labussière. 2009. "Wind Power Planning in France (Aveyron), from State Regulation to Local Planning." *Land Use Policy* 26 (3): 747.
9. Terral, Pierre-Marie. 2011. *Larzac: De La Lutte Paysanne à l'altermondialisme*. Paris: Privat.
10. See Sullivan, Sian. 2005. "'We Are Heartbroken and Furious!' Rethinking Violence and the (Anti-)Globalisation Movements." In: Catherine Eschle and Bice Maiguashca (eds), *Critical Theories, International Relations and the Anti-Globalisation Movement: The Politics of Global Resistance*. London: Routledge, pp. 175–194; Graeber, David. 2017. *Direct Action: An Ethnography*. Oakland: AK press, 2009.
11. See Vidalou, Jean-Baptiste. *Être Forêts: Habiter Des Territoires En Lutte*. Paris: Zones; Quadruppani, Serge. 2018. *Le Monde Des Grands Projets et Ses Ennemis: Voyage Au Coeur Des Nouvelles Pratiques Révolutionnaires*. Paris: Éditions La Découverte.
12. Terral, *Larza*.
13. Gildea, Robert, and Andrew Tompkins. 2015. "The Transnational in the Local: The Larzac Plateau as a Site of Transnational Activism since 1970." *Journal of Contemporary History* 50 (3): 582.
14. Ibid., p. 584.
15. Ibid.
16. Sullivan, "We Are Heartbroken and Furious!".

17. Leila and Scacha. 2001. "The Anarchist Ethic in the Age of the Anti-Globalization Movement." *Killing King Abacus* 2: 1–20. https://archive.elephanteditions.net/library/killing-king-abacus-the-anarchist-ethic-in-the-age-of-the-anti-globalisation-movement.pdf
18. MTC (Mauvaise Troupe Collective). 2018. *The Zad and NoTAV: Territorial Struggles and the Making of a New Political Intelligence*. New York: Verso Books.
19. Ibid.
20. Crimethinc. 2019. "Reflections on the ZAD: Another History." *Crimethinc*. https://crimethinc.com/2019/04/23/reflections-on-the-zad-looking-back-a-year-after-the-evictions
21. Le Monde. 2015. "De Notre-Dame-Des-Landes à Sivens, La Carte de France Des Projets Contestés." *Le Monde*, www.lemonde.fr/les-decodeurs/visuel/2015/12/21/la-carte-de-france-des-projets-contestes_4836014_4355770.html
22. See Quadruppani, *Le Monde Des Grands Projets et Ses Ennemis*; Robert, Diane. 2018. "Social Movements Opposing Mega Projects: A Rhizome of Resistance to the Neoliberal Hydra?" https://documents.pub/document/social-movements-opposing-mega-has-deeply-affected-mega-projects-they-imply.html?page=3
23. See Renée Conan, translated by Vincent Caillou. 2010. *Women of Plogoff*. Berkelely: Ardent Press.
24. Personal Communications, March 2015.
25. See a short video history of anti-airport show down outside Tokyo. It has been speculated that this is the first visual documentation of Molotov cocktails being used in media footage against the police: See, https://www.youtube.com/watch?v=zJMB01iscMo
26. Interview 5, 2 April 2018.
27. Interview 14, 7 April 2018.
28. Interview 9, 5 May 2018.
29. ADN. 2019. "Sud-Aveyron. Le Parc Dévoile Son 'Plan Climat.'" Online: *Aveyron Digital News*. www.aveyrondigitalnews.fr/2019/02/11/sud-aveyron-le-parc-devoile-son-plan-climat/.
30. Interview 18, 21 April 2019.
31. Consultation video and transcript, 25 May 2018.
32. Ibid.
33. Zehner, Ozzie. 2012. *Green Illusions: The Dirty Secrets of Clean Energy and the Future of Environmentalism*. Lincoln: University of Nebraska Press; Dunlap, Alexander. 2018. "End the 'Green' Delusions: Industrial-Scale Renewable Energy Is Fossil Fuel+." *Verso Blog*. www.versobooks.com/blogs/3797-end-the-green-delusions-industrial-scale-renewable-energy-is-fossil-fuel; Aronoff, Kate, Alyssa Battistoni, Daniel Aldana Cohen, and Thea Riofrancos. 2019. *A Planet to Win: Why We Need a Green New Deal*. New York: Verso Books; Hickel, Jason. 2019. "The Limits of Clean Energy." *Foreign Policy*. https://foreignpolicy.com/2019/09/06/the-path-to-clean-

energy-will-be-very-dirty-climate-change-renewables/; Sovacool, Benjamin K, Andrew Hook, Mari Martiskainen, Andrea Brock, and Bruno Turnheim. 2020. "The Decarbonisation Divide: Contextualizing Landscapes of Low-Carbon Exploitation and Toxicity in Africa." *Global Environmental Change* 60: 1–19; Dunlap, Alexander. 2021. "Does Renewable Energy Exist? Fossil Fuel+ Technologies and the Search for Renewable Energy." In *A Critical Approach to the Social Acceptance of Renewable Energy Infrastructures—Going beyond Green Growth and Sustainability*, edited by Susana Batel and David Philipp Rudolph. London: Palgrave, pp. 83–102.

34. Dunlap, 2018.
35. IEA. 2021. "Minerals Used in Clean Energy Technologies Compared to Other Power Generation Sources." https://www.iea.org/data-and-statistics/charts/minerals-used-in-clean-energy-technologies-compared-to-other-power-generation-sources
36. Dunlap, Alexander. 2018. "'A Bureaucratic Trap:' Free, Prior and Informed Consent (FPIC) and Wind Energy Development in Juchitán, Mexico." *Capitalism Nature Socialism* 29 (4): 88–108; Leifsen, Esben, Maria-Therese Gustafsson, Maria A Guzmán-Gallegos, and Almut Schilling-Vacaflor. 2017. "New Mechanisms of Participation in Extractive Governance: Between Technologies of Governance and Resistance Work." *Third World Quarterly* 38 (5): 1043–1057.
37. Interview 31, 3 May 2019.
38. Esmailzadeh, Sedigheh, Mouloud Agajani Delavar, Ashraf Aleyassin, Sayyed Asghar Gholamian, and Amirmasoud Ahmadi. 2019. "Exposure to Electromagnetic Fields of High Voltage Overhead Power Lines and Female Infertility." *The International Journal of Occupational and Environmental Medicine* 10 (1): 11; Gervasi, Federico, Rossella Murtas, Adriano Decarli, and Antonio Giampiero Russo. 2019. "Residential Distance from High-Voltage Overhead Power Lines and Risk of Alzheimer's Dementia and Parkinson's Disease: A Population-Based Case-Control Study in a Metropolitan Area of Northern Italy." *International Journal of Epidemiology* 48 (6): 1949–1957.
39. Bakker, Roel H., Eja Pedersen, Godefridus Petrus van den Berg, Roy E. Stewart, W. Lok, and J. Bouma. 2012. "Impact of Wind Turbine Sound on Annoyance, Self-Reported Sleep Disturbance and Psychological Distress." *Science of the Total Environment* 425: 42–51.
40. Two living within the St. Victor and outside, in the Occitanie.
41. Interview 32, 3 May 2019.
42. The couple interviewed—both working for EDF/RTE—began arguing because their partner refused to explicitly relate a tumor behind their eye (that caused permanent blindness) to working and living around energy infrastructure. Interview 32, 3 May 2019.
43. Interview 27, 27 April 2019. EC. "Opinion on Potential Health Effects of Exposure to Electromagnetic Fields." Online: European Commission, 2015.

https://ec.europa.eu/health/scientific_committees/emerging/docs/scenihr_o_041.pdf

44. Ibid.
45. For more on this region, see Franquesa, Jaume. 2018. *Power Struggles: Dignity, Value, and the Renewable Energy Frontier in Spain*. Bloomington: Indiana University Press.
46. Ibid.
47. Bonneuil, Christophe, and Jean-Baptiste Fressoz. 2016. *The Shock of the Anthropocene: The Earth, History and Us*. New York: Verso Books, p. 101.
48. Interview 12, 7 April 2018.
49. OWD. 2012. "Global Energy Consumption: How Much Energy Does the World Consume?" *Our World in Data*. https://ourworldindata.org/energy-production-consumption.
50. Kar-Gupta, Sudip, and Susanna Twidale. 2019. "EDF Warns UK Nuclear Plant Could Cost Extra $3.6 Billion, See More Delays." *Reuters*. www.reuters.com/article/us-britain-nuclear-hinkley-edf/edf-warns-uk-nuclear-plant-could-cost-extra-3-6-billion-see-more-delays-idUSKBN1WA0T0. For more details also read: Sullivan, Sian. "After the Green Rush? Biodiversity Offsets, Uranium Power and The 'Calculus of Casualties' in Greening Growth." *Human Geography* 6 (1): 80–101.
51. Reuters. 2019. "France Asks EDF to Prepare to Build 6 EPR Reactors in 15 Years -Le Monde." *Reuters*, www.reuters.com/article/us-edf-nuclear-epr/france-asks-edf-to-prepare-to-build-6-epr-reactors-in-15-years-le-monde-idUSKBN1WT27T
52. Interview 17, 21 April 2019.
53. Interview 20, 22 April 2019.
54. Anonymous. 2018. "Sabotage d'éolienne Acte 3 La Montagne Noire." *Nantes Indy Media*. www.nantes.indymedia.org/articles/42788.
55. Bárcenas, Francisco López. 2016. "Los 'Talibanes Indígenas' y El Despojo Capitalista." *La Jornada*, http://www.jornada.unam.mx/2016/10/08/opinion/017a1pol
56. Interview 9, 5 April 2018.
57. Rios Edwin. 2023. "Atlanta Police Charge 23 with Domestic Terrorism amid 'Cop City' Week of Action." *The Guardian*. 6 March 2023. www.theguardian.com/us-news/2023/mar/06/atlanta-georgia-cop-city-protest
58. Interview 8, 5 April 2018.
59. Interview 19, 21 April 2019.
60. Ibid.
61. Interview 6, 3 April 2018.
62. Interview 5, 2 April 2018.
63. Watch the 20 min video documenting the evictions, actions and attempts at land reclamation, see here: https://vimeo.com/455002348?signup=true Vive gato pirate!

64. OEC. 2017. *Communiquéé du colléctif réégional Toutés Nos Enérgiés-Occitanié Environnémént.* www.ventdecolere.org/actualites/Bouriege-11-historique.pdf
65. Ibid.
66. Interview 22, 22 October 2020.
67. Biasotto, Larissa D, and Andreas Kindel. 2018. "Power Lines and Impacts on Biodiversity: A Systematic Review." *Environmental Impact Assessment Review* 71: 110–119; Tabassum-Abbasi, M. Premalatha, Tasneem Abbasi, and S.A. Abbasi. 2014. "Wind Energy: Increasing Deployment, Rising Environmental Concerns." *Renewable and Sustainable Energy Reviews* 31 (1): 270–828.
68. Interview 25, 23 October 2020.
69. Interview 24, 23 October 2020. The OEC (2017, p. 8) says "€5,010 per wind turbine."
70. Interview 22, 22 October 2020.
71. Ibid.
72. OEC.
73. Sections de protection et d'intervention, a crime-gang and anti-terrorism unit.
74. Interview 22, 22 October 2020.
75. For video links and photographs, see OEC, 2017.
76. Nuss-Girona, Sergi, Joan Vicente Rufí, and Guillem Canaleta. 2020. "50 Years of Environmental Activism in Girona, Catalonia: From Case Advocacy to Regional Planning." *Land* 9 (6): 172.
77. Nogué, Joan, and Stephanie Wilbrand. 2010. "Landscape, Territory, and Civil Society in Catalonia." *Environment and Planning D: Society and Space* 28 (4): 638–652.
78. Original: Interconexión eléctrica Francia-España.
79. Francos, P. Labra, S. Sanz Verdugo, H. Fernández Álvarez, S. Guyomarch, and J. Loncle. 2012. "INELFE—Europe's First Integrated Onshore HVDC Interconnection," 1–8. *IEEE.*
80. Interview 48, 18 November 2020.
81. Interview 52, 24 November 2020.
82. Interview 44, 14 November 2020.
83. Biasotto and Kindel, "Power Lines and Impacts on Biodiversity."
84. Orion, Tao. 2015. *Beyond the War on Invasive Species: A Permaculture Approach to Ecosystem Restoration.* White River Junction: Chelsea Green Publishing.
85. Interview 59, 2 December 2020.
86. Biasotto and Kindel, "Power Lines and Impacts on Biodiversity," p. 114.
87. Franquesa, 2018.
88. Interview 46, 18 November 2020.
89. Interview 47, 18 November 2020.
90. Interview 48, 18 November 2020.
91. Interview 60, 4 December 2020.

92. Tsagas, I. 2019. "Spain's Third Interconnection with Morocco Could Be Europe's Chance for African PV—or a Boost for Coal." Available at: www.pv-magazine.com/2019/02/20/spains-third-interconnection-with-morocco-could-be-europes-chance-for-african-pv-or-a-boost-for-coal/
93. RTE. 2012. "Celtic Interconnector: Interconnection Project Between France and Ireland." Available at: www.rte-france.com/en/projects/celtic-interconnector-interconnexion-between-france-ireland#Theproject
94. Boulakhbar, M., B. Lebrouhi, Tarik Kousksou, S. Smouh, A. Jamil, M. Maaroufi, and M. Zazi. 2020. "Towards a Large-Scale Integration of Renewable Energies in Morocco." *Journal of Energy Storage* 32: 101–111.
95. Ibid., p. 8.
96. Ibid.
97. Allan, Joanna, Mahmoud Lemaadel, and Hamza Lakhal. 2021. "Oppressive Energopolitics in Africa's Last Colony: Energy, Subjectivities, and Resistance." *Antipode*, p. 7.
98. Cantoni, Roberto, and Karen Rignall. 2019. "Kingdom of the Sun: A Critical, Multiscalar Analysis of Morocco's Solar Energy Strategy." *Energy Research & Social Science* 51: 20–31.
99. WSRW. 2021. "Greenwashing Occupation." *Western Sahara Resource Watch*, https://vest-sahara.s3.amazonaws.com/wsrw/feature-images/File/405/616014d0c1f1d_Greenwashing-occupation_web.pdf
100. Allan et al., "Oppressive Energopolitics in Africa's Last Colony," p. 14.
101. Ibid., p. 13.
102. Ibid., p. 7.
103. Shelley, Toby. 2004. *Endgame in the Western Sahara: What Future for Africa's Last Colony*. London: Zed Books.
104. Allan et al., "Oppressive Energopolitics in Africa's Last Colony."
105. House, Freedom. 2020. "Freedom in the World 2020: Morocco." *Freedom House*. https://freedomhouse.org/country/morocco/freedom-world/2020.
106. Interview 63, 29 April -04-2021.
107. Allan et al. "Oppressive Energopolitics in Africa's Last Colony."
108. Ibid.
109. Interview 56, 30 November 2020.
110. European Commission. "The European Green Deal." Online: European Union, p. 9. https://eur-lex.europa.eu/legal-content/EN/TXT/?qid=1576150542719&uri=COM%3A2019%3A640%3AFIN
111. European Commission. 2021. *EU Funding Possibilities in the Energy Sector*. https://ec.europa.eu/energy/funding-and-contracts/eu-funding-possibilities-in-the-energy-sector_en
112. Ibid.
113. European Commission. 2021. *Financing Energy Efficiency*. https://ec.europa.eu/energy/topics/energy-efficiency/financing-energy-efficiency_en?redir=1
114. EC. 2017. *State Aid: Commisson Approves Spanish Support Scheme for Renewable Electricity*. Brussels: European Commission. https://ec.europa.eu/info/

news/state-aid-commission-approves-support-scheme-energy-intensive-companies-spain-2021-jan-11_en
115. Märkle-Huß, Joscha, Stefan Feuerriegel, and Dirk Neumann. 2018. "Contract Durations in the Electricity Market: Causal Impact of 15 Min Trading on the EPEX SPOT Market." *Energy Economics* 69: 367–378.
116. EPEX SPOT. (2020). *Opening New Horizons: Annual Report 2019*. www.epexspot.com/sites/default/files/sites/catalogue/
117. Märkle-Huß et al., "Contract Durations in the Electricity Market," p. 367.
118. Ocker, Fabian, and Vincent Jaenisch. 2020. "The Way towards European Electricity Intraday Auctions—Status Quo and Future Developments." *Energy Policy* 145: 111731.
119. Ibid., p. 3.
120. European Commission. 2021. *Antitrust: Commission Opens Investigation Into Possible Anticompetitive Behaviour By The Power Exchange EPEX Spot*. https://ec.europa.eu/commission/presscorner/detail/en/ip_21_1523
121. Interview 21, 21 August 2020.
122. EPEX SPOT.
123. Märkle-Huß et al., "Contract Durations in the Electricity Market"; SE (SchneiderElectric). 2017. "Understanding Renewable Energy Certificates in Europe: The Polices and Principles." SchneiderElectric. http://globalsustain.org/publish/understanding-renewable-energy-certificates-in-europe.pdf
124. Interview 21, 21 August 2020.
125. Ibid.
126. Bolger, Meadhbh, Diego Marin, Adrien Tofighi-Niaki, and Louelle Seelmann. 2021. "'Green Mining' Is a Myth: The Case for Cutting EU Resource Consumption." Brussels: European Environmental Bureau Friends of the Earth Europe. https://friendsoftheearth.eu/wp-content/uploads/2021/09/Methodology-considerations-Annex-to-green-mining-is-a-myth.pdf
127. Ibid.
128. Klesty, Victoria. 2021. "Electric Cars Hit 65% of Norway Sales as Tesla Grabs Overall Pole." *Reuters*. www.reuters.com/business/autos-transportation/electric-cars-take-two-thirds-norway-car-market-led-by-tesla-2022-01-03/

CHAPTER 6: WHEN ENVIRONMENTALISM IS ECOCIDE

1. Dunlap, Alexander. 2021. "Employing EU Public Money to Persuade Environmental Sacrifice: This Must End." *Yes to Life, No to Mining (YLNM) Network and European Parliament*, https://yestolifenotomining.org/wp-content/uploads/2021/12/Dr-Dunlap-full-testimony.pdf
2. FAO, 2020. "Globally Important Agricultural Heritage Systems—Barroso Agro-Sylvo-Pastoral System." *Food and Agricultural Organization of the United Nations*. www.fao.org/giahs/giahsaroundtheworld/designated-sites/europe-and-central-asia/barroso-agro-slyvo-pastoral-system/detailed-information/en/

3. While plantation agriculture is under discussion here, see: Lahiri-Dutt, Kuntala. 2018. "Extractive Peasants: Reframing Informal Artisanal and Small-Scale Mining Debates." *Third World Quarterly*: 1–22; Peluso, Nancy Lee. 2017. "Plantations and Mines: Resource Frontiers and the Politics of the Smallholder Slot." *The Journal of Peasant Studies* 44 (4): 834–869.
4. Bolger, Meadhbh, Diego Marin, Adrien Tofighi-Niaki, and Louelle Seelmann. 2021. "'Green Mining' Is a Myth: The Case for Cutting EU Resource Consumption." Brussels: European Environmental Bureau Friends of the Earth Europe. https://friendsoftheearth.eu/wp-content/uploads/2021/09/Methodology-considerations-Annex-to-green-mining-is-a-myth.pdf
5. Demony, Catarina. 2023. "Portugal Gives Environmental Green Light for Savannah's Lithium Mine." *Reuters*, www.reuters.com/markets/commodities/savannah-says-portugal-gives-environmental-green-light-lithium-mine-2023-05-31/
6. EC, 2020. "Critical Raw Materials For Strategic Technologies and Sectors In The EU—A Foresight Study." *European Commission*, https://ec.europa. eu/docsroom/documents/42881
7. Ibid.
8. Bridge, G., and E. Faigen, 2022. "Towards the Lithium-Ion Battery Production Network: Thinking Beyond Mineral Supply Chains." *Energy Res. Soc. Sci.*: 13
9. EC, "Critical Raw Materials For Strategic Technologies and Sectors In The EU"; Dunlap, J. 2014. "Fairhead, The Militarisation and Marketisation of Nature: An Alternative Lens to 'Climate-Conflict.'" *Geopolitics*, 19 (4): 937–961.
10. Northvolt. "Galp and Northvolt Select Setúbal to Build Advanced Lithium Conversion Unit." *Northvolt*, https://northvolt.com/articles/aurora-setubal/
11. Bolger et al., "'Green Mining' is a Myth," p. 3.
12. Klesty, V. 2021. "Electric Cars Hit 65% Of Norway Sales As Tesla Grabs Overall Pole." *Reuters*. www.reuters.com/business/autos-transportation/electric-cars-take-two-thirds-norway-car-market-led-by-tesla-2022-01-03/
13. EC. 2019. "Clean Energy for All Europeans. Available at European Union." https://op.europa.eu/en/publication-detail/-/publication/b4e46873-7528-11e9-9f05-01aa75ed71a1/language-en?WT.mc_id=Searchresult&WT.ria_c=null&WT.ria_f=3608&WT.ria_ev=search EC, 2020.
14. Carvalho, A., M. Riquito, V. Ferreira. 2022. "Sociotechnical imaginaries of energy transition: the case of the Portuguese Roadmap for Carbon Neutrality 2050." *Energy Rep.*: 82413–82423.
15. Charoy, B., F. Lhote, Y. Dusausoy et al., 1992. "The Crystal Chemistry Of Spodumene In Some Granitic Aplite-Pegmatite Bodies of Northern Portugal; A Comparative Review. *Can. Mineral.* 30 (3): 639–651.
16. These institutions have changed their names over the course of the years: Instituto Geológico e Mineiro (1993–2003); Instituto Nacional de Engenha-

ria e Tecnologia e Industrial (2003–2006); Laboratório Nacional de Energia e Geologia (LNEG) (2006–present).

17. República Portuguesa. 2017. *Relatório do grupo de trabalho "lítio." despacho nº 15040/2016 de SEE.* Lisboa: March 2017.
18. Carballo-Cruz, F., and J. Cerejeira. 2020. "The Mina do Barroso Project—Economic and Development Impacts." *University of Minho*. www.savannahresources.com/media/uuri54jx/the-mina-do-barroso-project-economic-development-impacts_universityofminho_english_final.pdf. This recurring statistic comes from a study commissioned by Savannah Resources, done by researchers from the Minho University on the economic impacts and development of the "Mina do Barroso" project.
19. Gomes, M.E.P., and J.M.F. Ramos. 2018. *Recursos minerais de Trás-os-Montes e Alto Douro C. Balsa.* In: J.S. Teixeira (Eds.), Recursos geológicos de Trás-os-Montes—Passado, Presente e perspetivas futuras, Instituto Politécnico de Bragança, Bragança (2018), p. 40.
20. USGS. 2022. "Mineral Commodities Summaries 2022." *United States Geological Survey*. https://pubs.usgs.gov/periodicals/mcs2022/mcs2022-lithium.pdf
21. BGEN. "Portugal to Call Off Lithium Project amid EU's Scramble For Battery Materials." *Balkan Green Energy News*, https://balkangreenenergynews.com/portugal-to-call-off-lithium-project-amid-eus-scramble-for-battery-materials/
22. República Portuguesa, 2017.
23. QUERCUS. 2019. *Levantamento dos pedidos de prospeção e pesquisa de depósitos minerais (2016–2019)*. QUERCUS-ACN.
24. Repulica Portuguesa, 2017.
25. MWP, Mapa do Minério. 2022. "Mining Watch Portugal." https://mining-watch.pt/mapadominerio/index.html
26. Gomes and Ramos, *Recursos minerais de Trás-os-Montes e Alto Douro C. Balsa.* Gomes, C.L.A. 2018. *Panorâmica sobre condições naturais de ocorrência de minérios de lítio no Norte de Portugal—Perspectivas de valorização de recursos de Lítio metálico.* C. Balsa and J.S. Teixeira (Eds.), Recursos geológicos de Trás-os-Montes—Passado, Presente e perspetivas futuras, Instituto Politécnico de Bragança, Bragança (2018).
27. Fagundes, T. 2021. "O regress do fantasma das minas da Borralha." *Jornal MAPA*. www.jornalmapa.pt/2021/03/30/o-regresso-do-fantasma-das-minas-da-borralha/ (2021).
28. Ávila, P.F., S. Vieira, C. Candeias, E. Ferreira da Silva, 2015. "Assessing heavy metal/metalloids pollution in soils after eight decades of intense mining exploration—the case study of Borralha mine." *Portugal Comun.Geol.* 102: 57–61, Ribeiro, J.I.V. 2010. *Levantamento do estado de contaminação de solos e águas superficiais da antiga Mina da Borralha.* Master Thesis. Mines and Geo-Environment Engineering; Faculty of Engineering; Porto University. https://repositorio-aberto.up.pt/handle/10216/59270; Vieira, S. 2014.

"Risco associado à exposição a teores elevados de metais na área mineira da Borralha." Geosciences Department, University of Aveiro.

29. Dhar, A., M.A. Naeth, P.D. Jennings et al. 2020. Perspectives on Environmental Impacts and a Land Reclamation Strategy For Solar And Wind Energy Systems." *Sci. Total Environ.* 718: 1–9.
30. Chaves, C., E. Pereira, P. Ferreira et al. 2021. "Concerns about Lithium Extraction: A Review and Application for Portugal." *Extractive Industries and Society* 8 (3): Article 100928.
31. Emerman, S.H. 2021. "Testimony of Dr. Steven H. Emerman to the European Parliament Public Hearing on Environmental and Social Impacts of Mining in the EU." *Yes to Life, No to Mining/European Commission.* https://yestolifenotomining.org/wp-content/uploads/2021/12/Prof-Emerman-Testimony.pdf
32. Field notes.
33. DGEG, 2011. "Aviso 8931/2011, de 13 de Abril. Available at Direção-Geral de Energia e Geologia." https://dre.tretas.org/dre/1241101/aviso-8931-2011-de-13-de-abril
34. Emerman, "Testimony of Dr. Steven H. Emerman."
35. VISA Consultores. 2021. "Environmental Impact Study. Non-Technical Resume. Expansion of the Barroso Mine."
36. Emerman, "Testimony of Dr. Steven H. Emerman," p. 16.
37. Ibid.
38. Ibid.
39. Savannah. 2021. "Mina do Barroso EIA Update: Portuguese Environment Agency Declares Conformity of EIA." *Savannah Resources.* www.savannahresources.com/investors/rns-feed/rns-announcements/?rid=4019404
40. Savannah. 2021. "Environmental Impact Study: Non-Technical Resume." *Savannah Resources.* www.savannahresources.com/media/crvdaoeo/ntr-of-mdb-april-2021.pdf
41. Jornal de Negócios. 2022. APA adia decisão sobre mina de lítio do Barroso para março de 2023. *Jornal de negócios.* www.jornaldenegocios.pt/economia/ambiente/detalhe/apa-adia-decisao-sobre-mina-de-litio-do-barroso-para-marco-de-2023
42. Field notes.
43. FAO, 2020, "Globally Important Agricultural Heritage Systems."
44. Maravalhas, E., J.M. Arantes, A. Maravalhas. 2022. "Biodiversidade do Barroso. Boticas Parque, Natureza e Biodiversidade, Boticas."
45. FAO. 2018. "Globally Important Agricultural Heritage Systems." S*istema Agro-silvo-pastoril do Barroso, Portugal,* www.fao.org/giahs/giahsaroundtheworld/designated-sites/europe-and-central-asia/barroso-agro-slyvo-pastoral-system/pt/ Accessed 11 December 2022.
46. E. Maravalhas et al.
47. FAO, 2020, "Globally Important Agricultural Heritage Systems."

48. FAO, 2020, "Globally Important Agricultural Heritage Systems."
49. FAO, 2018, "Globally Important Agricultural Heritage Systems."
50. Field notes and Interview 16, 21 January 2022.
51. VISA Consultores, "Environmental Impact Study."
52. All the citizens who reside in the area where the "baldios" are located are automatically "compartes," in correspondence with the habits and customs recognized by the Local Communities. Non-resident citizens can also be granted the status of "compartes" by the Assembly of the Compartes.
53. Dunlap, A. 2020. "Bureaucratic Land Grabbing For Infrastructural Colonization: Renewable energy, L'Amassada and Resistance in Southern France." *Human Geography*, 13 (2): 109–126; Dunlap, A. "Spreading 'Green' Infrastructural Harm: Mapping Conflicts And Socioecological Disruptions Within the European Union's Transnational Energy Grid." *Globalizations* 202: 1–25. https://doi.org/10.1080/14747731.2021.1996518. p. 13.
54. Interview 7, 20 January 2022.
55. Interview 26, 30 May 2022.
56. Interview 6, 19 Janaury 2022.
57. Interview 1, 16 January 2022.
58. Savannah. 2021. "Environmental Impact Study: Non-Technical Resume." *Savannah Resources*. www.savannahresources.com/media/crvdaoeo/ntr-of-mdb-april-2021.pdf
59. Interview 15, 21 January 2022.
60. Ibid.
61. Field notes.
62. Interview 14, 21 January 2022.
63. Interview 15, 21 January 2022.
64. Field notes and Interview 15, 21 January 2022.
65. Interview 26, 30 May 2022.
66. Interview 4, 19 January 2022. Interview 7, 20 January 2022. Interview 15, 21 January 2022.
67. Interview 4, 19 January 2022.
68. Ibid.
69. Interview 15, 21 January 2022.
70. Ibid.
71. Interview 9, 20 January 2022.
72. Interview 19, 23 January 2022.
73. Ibid.
74. Interview 11, 20 January 2022.
75. Interview 4, 10 January 2022.
76. Interview 10, 20 January 2022.
77. Savannah, "Environmental Impact Study," pp. 33–34.
78. Interview 17, 22 January 2022.
79. Fagundes, T. 2022. "O regresso do fantasma das Minas da Borralha." *Journal MAPA*. www.jornalmapa.pt/2021/03/30/o-regresso-do-fantasma-das-minas-da-borralha/ Accessed 11 December 2022.

80. Field Notes.
81. Ibid.
82. Chaves et al., "Concerns about Lithium Extraction"; Brock, A. 2020. "Frack Off: Towards An Anarchist Political Ecology Critique Of Corporate and State Responses To Anti-Fracking Resistance in the UK." *Political Geography*: 1–15; Conde, M., P. Le Billon. 2017. "Why Do Some Communities Resist Mining Projects While Others Do Not?" *The Extractive Industries and Society* 4 (3): 681–697; Jakobsen, L.J. 2020. "Corporate Security Technologies: Managing Life And Death Along a Colombian Coal Railway." *Polit. Geogr.* 83: 1–10; Dunlap, A. 2020. "Wind, Coal, and Copper: The Politics of Land Grabbing, Counterinsurgency, and the Social Engineering of Extraction." *Globalizations* 17 (4): 661–682.
83. Interview 20, 24 January 2022.
84. Interview 7, 20 January 2022.
85. Interview 17, 22 January 2022.
86. Savannah, "Environmental Impact Study," p. 15.
87. Bridge and Faigen, "Towards the Lithium-Ion Battery Production Network"; Chaves et al., "Concerns about Lithium Extraction."
88. WISE. 2022. "Chronology of Major Tailings Dam Failure: from 1960 to 2022." *Wise Uranium Project*. www.wise-uranium.org/mdaf.html (2022)
89. Conde and Le Billon, "Why Do Some Communities Resist Mining Projects While Others Do Not?"; Le Billon P. and M. Sommerville. 2018. "Landing Capital and Assembling 'Investable Land' in the Extractive and Agricultural Sectors." *Geoforum* 82: 212–224; Kirsch, S. 2014. *Mining Capitalism: The Relationship Between Corporations and Their Critics*. Berkeley: University of California Press; Gilbert, J.E., T. Gilbertson, and L. Jakobsen. 2021. "Incommensurability and Corporate Social Technologies: A Critique Of Corporate Compensations in Colombia's Coal Mining Region of La Guajira." *J. Pol. Ecol.*, 28 (1): 434–452.
90. Interview 21, 24 January 2022.
91. Savannah, "Environmental Impact Study."
92. Emerman, "Testimony of Dr. Steven H. Emerman."
93. Savannah, "Environmental Impact Study."
94. P. Le Billon and M. Sommerville, "Why Do Some Communities Resist Mining Projects While Others Do Not?"
95. Fieldnotes and Interview 24, 25 January 2022.
96. Interview 16, 21 January 2022.
97. Simpson, M. "Resource Desiring Machines: The Production of Settler Colonial Space, Violence, and the Making of a Resource in the Athabasca Tar Sands." *Political Geography* 74: 1–12; Scott, J.C. 1998. *Seeing Like a State: How Certain Schemes to Improve the Human Condition Have Failed*. New Haven: Yale University Press.
98. Dunlap, A. 2019. *Renewing Destruction: Wind Energy Development, Conflict and Resistance in a Latin American Context*. London: Rowman & Littlefield.

99. Interview 24, 25 January 2022 and Field Notes.
100. Aroso, J.S. and O. Magalhães. 2012. "The Mining Law Review: Portugal." https://thelawreviews.co.uk/title/the-mining-law-review/portugal-mining-law (2021)
101. "IRC is levied at a 21% rate, to which may be added a municipal surtax of up to 1.5% levied on taxable profits (depending on the municipality), as well as a state surtax of 3% on taxable profits exceeding €1.5 million and up to €7.5 million, 5% on taxable profits exceeding €7.5 million and up to €35 million, and 9% on taxable profits exceeding €35 million. This means that the effective tax rate can reach 22.5%, to which will be added the state surtax, to which the above rates are applied in a staggered way. A special reduced IRC rate (of 17% on taxable profits up to €25,000) is available for small and medium companies (with a turnover below €50 million among other criteria established by law)." RTP. 2020. Câmara de Montalegre. Autarca arrisca-se a perder mandato. *RTP Notícias*, p. 4. https:// mandato_v1269677
102. Savannah, "Environmental Impact Study."
103. Interview 7, 20 January 2022.
104. Interview 19, 23 January 2022. The interviewee avoids giving specifics on the number of jobs by talking about the "lack of licensed staff" (e.g., skilled workers) and "lack of unlicensed staff" (unskilled workers). The answer to this question is interpreted as trying to say something "intelligent" without giving specifics.
105. Interview 19, 23 January 2022 and Field Notes.
106. 106 Savannah, "Environmental Impact Study."
107. Interview 15, 21 January 2022.
108. Field notes.
109. Interview 20, 24 January 2022.
110. Interview 5.
111. Interview 2 19 January 2022.
112. Interview 18, 22 January 2022.
113. Interview 12, 21 January 2022.
114. Interview 15, 21 January 202.
115. Interview 8, 20 January 2022.
116. Interview 20, 24 January 2022.
117. Interview 16, 21 January 2022.
118. Interview 20, 24 January 2022.
119. Interview 8, 20 January 2022. Interview 17, 22 January 2022.
120. Interview 17, 22 January 2022.
121. Interview 20, 24 January 2022.
122. RTP. 2020. Câmara de Montalegre. Autarca arrisca-se a perder mandato. *RTP Notícias*. https:// mandato_v1269677
123. Rosa, L. and T. Pereirinha. 2022. "Socialista Orlando Alves, autarca de Montalegre, detido por suspeitas de associação criminosa, abuso de poder e recebimento indevido." *Observador*. https://observador.pt/2022/10/27/presidente-da-camara-de-montalegre-detido/

124. Field Notes. Interview 20, 24 January 2022. More about this on the Portuguese investigative media program "Sexta às 9."
125. Interview 5, 19 January 2022.
126. Lusa, M.P. 2017. "Abre inquérito ao caso de suspeita de fraude eleitoral em Montalegre." *Diário de Notícias*, www.dn.pt/portugal/autarquicas-mp-abre-inquerito-ao-caso-de-suspeita-de-fraude-eleitoral-em-montalegre-8849959.html
127. "There is no difference between an open-pit quarry and an open-pit lithium mine," said João Galamba, the Secretary of State for Energy, on 14 November 2019, on national television, while being interviewed for the program "Negócios da Semana" (SIC Notícias). Teresa Ponce de Leão, the President of LNEG, repeated the same sentence on 16 February 2022, on national television, during the program "Fronteiras XXI" (RTP 1).
128. Interview 17, 22 January 2022.
129. Escobar, A. 2012. *Encountering Development: The Making and Unmaking of the Third World*. Princeton: Princeton University Press; Klein, E. and C.E. Morreo, 2019. *Postdevelopment in Practice: Alternatives, Economies, Ontologies*. London: Routledge; Rahnema, M. and V. Bawtree 1997. *The Post-Development Reader*. London: Zed Books.
130. Interview 20, 21 January 2022.
131. Interview 5, 19 January 2022.
132. Interview 20 and 25.
133. Interview 9, 20 January 2022.
134. Interview 1, 19 January 2022.
135. Field notes.
136. Interview 4, 19 January 2022.
137. Interview 16, 21 January 2022.
138. Interview 20, 21 January 2022.
139. Gelderloos, P. 2022. *The Solutions Are Already Here: Tactics for Ecological Revolution From Below*. London: Pluto Books.
140. Interview 16, 21 January 2022.
141. Interview 7, 20 January 2022.
142. Interview 1, 20 January 2022.
143. Interview 7, 20 January 2022.
144. Interview 10, 20 January 2022.
145. Interview 10, 20 January 2022.
146. Interview 10, 20 January 2022.
147. Interview 10, 20 January 2022.
148. P. Le Billon, M. Sommerville, 2017.
149. Batel, S. and S. Küpers. 2022. "Politicizing Hydroelectric Power Plants in Portugal: Spatio-Temporal Injustices and Psychosocial Impacts of Renewable Energy Colonialism in the Global North." *Globalizations*: 1–20.
150. Interview 16, 21 January 2022.

151. Demony, Catarina. 2023. "Portugal Gives Environmental Green Light for Savannah's Lithium Mine." *Reuters*. www.reuters.com/markets/commodities/savannah-says-portugal-gives-environmental-green-light-lithium-mine-2023-05-31/
152. Gelderloos, Peter. 2017. *Worshiping Power: An Anarchist View of Early State Formation*. Oakland: AK Press.
153. Reinert, H. 2018. "Notes from a Projected Sacrifice Zone." *ACME* 17 (2): 597–617.
154. Velicu, I. 2020. "Prospective Environmental Injustice: Insights From Anti-Mining Struggles in Romania and Bulgaria." *Environ. Polit.* 29 (3): 414–434.
155. Batel and Küpers, "Politicizing Hydroelectric Power Plants in Portugal."

CONCLUSION

1. Fanon, Frantz. 1963. *The Wretched of the Earth*. Translated by Constance Farrington. New York: Grove Press; Rodney, Walter. 1972. *How Europe Underdeveloped Africa*. Washington DC: Howard University Press; Galeano, Eduardo. 1997. *Open Veins of Latin America: Five Centuries of The Pillage of a Continent*. London: Monthly Review Press; Moses, A. D., ed. 2008. *Empire, Colony, Genocide: Conquest, Occupation, and Subaltern Resistance in World History*. War and Genocide. Oxford: Berghahn.
2. See Bookchin, Debbie. 2017. "Radical Municipalism: The Future We Deserve." *Roar*. https://roarmag.org/magazine/debbie-bookchin-municipalism-rebel-cities/; Bookchin, Murray. 1991. "Libertarian Municipalism: An Overview." *Green Perspectives* 24: 1–6.
3. Tarinski, Tarinski Yavor. 2021. *Enlightenment and Ecology: The Legacy of Murray Bookchin in the 21st Century*. London: Black Rose Books Ltd.
4. Thompson, Matthew. 2021. "What's so New about New Municipalism?" *Progress in Human Geography* 45 (2): 317–342.
5. Sale, Kirkpatrick. 2000. *Dwellers in the Land: The Bioregional Vision*. Geogia: University of Georgia Press.
6. Gutierrez-Aguilar, Raquel. 2014. "Beyond the 'Capacity to Veto': Reflections from Latin America on the Production and Reproduction of the Common." *South Atlantic Quarterly* 113 (2): 259–270; Esteva, Gustavo. 2014. "Commoning in the New Society." *Community Development Journal* 49 (suppl_1): 144–159; García-López, Gustavo A., Ursula Lang, and Neera Singh. 2012. "Commons, Commoning and Co-Becoming: Nurturing Life-in-Common and Post-Capitalist Futures (An Introduction to the Theme Issue)." *Environment and Planning E: Nature and Space* 4 (4): 1199–1216.
7. Zapatistas. 2016. *Critical Thought in the Face of the Capitalist Hydra I*. Contributions by the Sixth Commission of the EZLN. Durham: Paper Boat Press; Rosset, Peter M., and Lia Pinheiro Barbosa. 2021. "Peasant Autonomy: The Necessary Debate in Latin America." *Interface: A Journal on Social Movements* 13 (1): 46–89.

8. Kothari, Ashish, Ariel Salleh, Arturo Escobar, Federico Demaria, and Alberto Acosta. 2019. *Pluriverse: A Post-Development Dictionary*. Delhi: University of Colombia Press.
9. Altmann, Philipp. 2020. "The Commons as Colonisation—The Well-Intentioned Appropriation of Buen Vivir." *Bulletin of Latin American Research* 39 (1): 83–97.
10. Ibid.
11. Gelderloos, Peter. 2022. *The Solutions Are Already Here: Tactics for Ecological Revolution From Below*. London: Pluto, pp. 173–204.
12. For a typology of tactics see: Sovacool, Benjamin K, and Alexander Dunlap. 2022. "Anarchy, War, or Revolt? Radical Perspectives for Climate Protection, Insurgency and Civil Disobedience in a Low-Carbon Era." *Energy Research & Social Science* 86: 1–17.
13. Emphasis added; IGD (It's Going Down). 2023. "'The Amount of Solidarity Is Incredible Here': Voices on the Frontlines of the Fight to Stop Cop City." *It's Going Down*, 6 March 2023. https://itsgoingdown.org/solidarity-incredible-here-atl-cop-city/
14. Sachs, Wolfgang, eds. 1992. *The Development Dictionary: A Guide to Knowledge as Power*. London: Zed Books; Rahnema, Majid, and Victoria Bawtree, eds. 1997. *The Post-Development Reader*. London: Zed Books; Escobar, Arturo. 2012. *Encountering Development: The Making and Unmaking of the Third World*. Princeton: Princeton University Press; Esteva, Gustavo. 2022. *Gustavo Esteva: A Critique of Development and Other Essays*. New York: Routledge.
15. Rahnema and Bawtree; Churchill, Ward. 2001. *A Little Matter of Genocide: Holocaust and Denial in the Americas 1492 to the Present*. San Francisco: City Lights Publisher. Levene, Mark. 2008. "Empires, Native Peoples, and Genocide." In *Empire, Colony, Genocide: Conquest, Occupation, and Subaltern Resistance in World History*, edited by A.D. Moses. Oxford: Berghahn. Moses, A.D., ed. 2008. *Empire, Colony, Genocide: Conquest, Occupation, and Subaltern Resistance in World History*. War and Genocide. Oxford: Berghahn; Moses, Dirk, and Dan Stone.2013. *Colonialism and Genocide*. London: Routledge; Short, Damien. 2010. "Cultural Genocide and Indigenous Peoples: A Sociological Approach." *The International Journal of Human Rights* 14 (6): 833–848; Woolford, Andrew, Jeff Benvenuto, and Alexander Laban Hinton, eds. 2014. *Colonial Genocide in Indigenous North America*. Durham: Duke University Press.
16. For more details read: Illich, 1978; Sachs, 1992; Escorbar, 1995/2012; Rahnema and Bawtree, 1997; Kothari and colleagues, 2019; Esteva, 2022.
17. van der Walt, L. 2018. "Anarchism and Marxism." In: N. Jun (ed), *The Brill Companion to Anarchism and Philosophy*. Leiden: Brill, 535.
18. Dunlap, Alexander, and Jostein Jakobsen. 2020. *The Violent Technologies of Extraction: Political Ecology, Critical Agrarian Studies and The Capitalist Worldeater*. London: Palgrave.

Index

fig refers to a figure; *ill* to an illustration